ANTHONY SMITH

Bakery:
Bread & Fermented Goods

Books by
L. J. HANNEMAN

Modern Cake Decoration
Cake Design and Decoration
(with G. Marshall)
Patisserie
Bakery: Bread & Fermented Goods
Bakery: Flour Confectionery
Gateaux and Torten

Bakery: Bread & Fermented Goods

L. J. HANNEMAN
F.Inst.B.B., F.H.C.I.M.A., F.C.F.A.(C.G.)

Formerly Head of Department of Service Industries Lancaster and Morecambe College of Further Education

HEINEMANN : LONDON

William Heinemann Ltd
10 Upper Grosvenor Street, London W1X 9PA
LONDON MELBOURNE
JOHANNESBURG AUCKLAND

First published 1980
Reprinted 1986

ISBN 0 434 90708 1

Filmset by Wilmer Brothers Limited, Birkenhead, Merseyside
Printed in Great Britain by REDWOOD BURN LIMITED, Trowbridge, Wiltshire

Preface

This is the first volume of a two volume work covering most aspects of the craft of Bakery and Flour Confectionery as follows – BOOK 1 Bread & Fermented goods; BOOK 2 Flour Confectionery.

This volume is essentially a book written for bakers who are interested in maintaining and increasing good standards of craft within our Industry. There will always be customers for bakery goods which are different from those produced by the large manufacturer and sold in every supermarket in the country. However, this publication is not only written for the craftsman baker, but also the ambitious amateur who may wish to bake in his own home. Teachers of Home Economics will find it a most valuable reference book.

This book has also been especially written for students who, it is hoped, will find within its pages most of the subject matter covered by the examination syllabus of the *City and Guilds of London Institute* 120 parts I, II and III.

Every effort has been made to ensure that the book contains the latest information on regulations and legislation relating to various baked products in Britain.

QUANTITIES USED IN THIS BOOK

All weights, volumes and measurements used in this book are expressed in both Metric and Imperial Units. Recipe proportions have been dictated by taking 1 kilogram ($2\frac{1}{4}$ lb) of the main basic ingredient (usually flour).

One great advantage of using the kilogram (1000 grams) as the main weight unit in recipes, is that any calculation involving the percentage of an ingredient based upon the flour weight becomes easy to establish in either metric or imperial units. A typical example using a bread recipe should suffice to explain this fact.

	Metric (grams)	Imperial (ozs)	% based upon flour weight %
Flour	1000	1000	100
Water	570	570	57
Yeast	30	30	3
Salt	20	20	2
Totals	1620	1620	162.0

Suppose we require a recipe in imperial units which will give us 5 lb of bread dough = 80 ounces, we would require:

	Calculation	Actual weight in ounces	Weight used in Practice
Flour	$\frac{80}{1620} \times \frac{1000}{1} =$	49·38	3 lb $1\frac{1}{4}$ oz
Water	$\frac{80}{1620} \times \frac{570}{1} =$	28·14	1 lb $12\frac{1}{4}$ oz
Yeast	$\frac{80}{1620} \times \frac{30}{1} =$	1·48	$1\frac{1}{2}$ oz
Salt	$\frac{80}{1620} \times \frac{20}{1} =$	·98	1 oz.
	Totals	79·98 oz	5 lb 0 oz

So a recipe which will give us exactly 5 lb of bread dough will now read:

Flour	3 lb $1\frac{1}{4}$ oz
Water	1 lb $12\frac{1}{4}$ oz
Yeast	$1\frac{1}{2}$ oz
Salt	1 oz
Total	5 lb 0 oz

A recipe for any given quantity in metric units can be calculated in the same way. For exact equivalents refer to the tables in the Appendices.

Acknowledgements

My grateful thanks are given to all who have helped me in the preparation of this book. In particular I wish to record my appreciation to the following:

The Principal and Governors of the Lancaster & Morecambe College of Further Education, who permitted me facilities of the College to be used for making many of the examples of the work illustrated; my colleagues, Arthur Green and Ray Ward, for their assistance and helpful advice, and Donald Bland for allowing the use of his bakery to produce some of the examples shown; the Flour Advisory Bureau for permission to reproduce the diagram on page 45, and the table on page 6; the City of Canterbury Environmental Health Department for permission to reproduce a large proportion of their handbook *Thought For Food* in the Food Hygiene section of the chapter "Legislation"; Consumers' Association for permission to reproduce fig. 47 on page 166, from their magazine *Which*?; Food, Drink and Tobacco Industrial Training Board for the reproduction of the letter on page 167 from their *Guide to the Health & Safety at Work Act*; the H.M.S.O. for permission to use extracts from their publication *Health and Safety at Work* in the chapter on "Legislation".

Finally I wish to dedicate this publication and acknowledge the debt I owe to those many craftsmen and craftswomen, past and present, who, without having given so freely of their knowledge, this book could not have been written.

Contents

1. Raw Materials

The quality of any bakery goods made depends as much upon the choice of the raw materials as the skills employed in the manufacture of the products. Therefore some notes on the raw materials available to the baker and confectioner are given here so that the right choice can be made for any particular product.

Details of raw materials will be grouped under general headings alphabetically for easy reference, (e.g. wheat flour is listed under cereals).

ACIDS

Acetic (Ethanoic)

An organic acid contained in vinegar to the extent of 3–6%. In the Baking Industry it has two uses.

(*a*) As a preventative against the disease known as rope in bread (*see* page 59).
(*b*) As an addition to royal icing to help it set and dry more rapidly.

Acid Calcium Phosphate (**A.C.P.**)

This has two uses.

(*a*) As an ingredient of baking powder
(*b*) A preventative for the disease known as rope (*see* page 59).

Ascorbic

Vitamin C used as an oxidant for the mechanical dough development system of bread making, i.e. Chorleywood Bread Process (*see* page 53).

Citric (2-hydroxy propane tricarboxylic)

Found naturally in fresh fruits but can be obtained as a white crystalline powder. It can be used as follows:

(*a*) To set jellies containing pectin.
(*b*) Acidulate sugar syrups to inhibit crystallization.
(*c*) When a fruit flavour is added to a filling cream, citric acid may be added to simulate the acid which would have been present if the natural fruit had been used, i.e. lemon.

Cream of Tartar (Argol) (Potassium hydrogen dihydroxybutanedioate)

A white crystalline powder which is the hydrogen potassium salt of tartaric acid (*see* below). It has several uses as follows:

(*a*) Ingredient of baking powder.
(*b*) To acidulate sugar syrups to help inversion and prevent crystallization.
(*c*) Added to fruit cake batters to help prevent the sinking of fruit.
(*d*) It is sometimes used in puff pastry in order to confer a measure of extra extensibility on the gluten of the flour.

Cream Powders

In organic acids such as sodium pyrophosphate or mixtures of this with others used as the alternative acid for mixing with sodium bicarbonate for making baking powder.

Glucono Delta Lactone (G.D.L.)

A white crystalline powder especially manufactured for compounding baking powder.

Proprionic Acid

Used for the control of rope (*see* page 59).

Sodium Pyrophosphate

Used for the control of rope (*see* page 59).

Sorbic Acid

Used for the control of mould (*see* page 60).

Tartaric Acid (Dihydroxybutaredioc)

This is very similar to its salt cream of tartar but more soluble in cold water. Because it is contained in all fruits it can therefore be used as an alternative to citric acid. It can also be used as a more acceptable alternative to cream of tartar for uses (b) (c) and (d).

BAKING POWDER

Applies to any substance or mixture which when moistened and heated will produce a gas which will aerate flour confectionery goods. This can easily be made by the baker or flour confectioner by mixing the correct proportions of sodium bicarbonate and a recommended acid in the proportions of 1 of sodium bicarbonate and 2 of acid.

Most baking powders available for sale are made up from mixtures of sodium bicarbonate, a suitable acid and a filler such as starch.

Because they react in the presence of water, they should always be stored in a dry place and used fresh.

Vol

This is ammonium carbonate which when moistened and heated produces ammonia gas, carbon dioxide and water.

Bicarbonate of Soda (Baking Soda)

A white crystalline powder which has three uses in Flour Confectionery:

(*a*) Ingredient in baking powder.
(*b*) Aerating agent in ginger goods.
(*c*) Improves the colour of chocolate cake when cocoa powder is used.

BLACKJACK (Caramel)

A thick syrup made from burnt sugar and used for colouring purposes, particularly celebration cakes such as Wedding, Christmas etc. It has an indefinite storage life.

CEREALS

General

All cereal products should be stored with care in a cool dry place and kept only for a limited period of time, otherwise they develop off flavours. They should also be stored in such a way that they are prevented from contamination from insects and rodents.

BARLEY

Milled into flour, this can be blended with wheaten flour and made into bread and flour confectionery goods. It has a low gluten content and therefore cannot be used on its own for breadmaking purposes. The main use of this cereal is for the production of *malt*.

MAIZE

This is milled and processed to form a white powder called *cornflour* which has several uses in flour confectionery as follows:

(*a*) Added to a strong flour to dilute its strength.
(*b*) Dusting purposes e.g. sugar paste.
(*c*) Ingredient in certain recipes e.g. sand cakes.
(*d*) Ingredient in various glazes, jellies and custards.
(*e*) Making of moulds for fondants.

In the manufacture of cornflour, most of the protein is removed and the product is practically pure starch.

However, maize may also be milled into a flour in which the protein is retained and this may be used to make corn-bread.

OATS

This is milled to give several types of products:

Oatmeal Coarse, Medium or Fine (pinhead) all of which may be used in various goods such as parkins and scones.

Rolled Oats The oats are partially cooked and are flattened by passing through rollers. These may be used for making flapjacks.

RICE

Milling produces two products:

Riceflour

This is a finely milled product used as follows:

(*a*) Ingredient in many types of products.
(*b*) Dusting purposes.
(*c*) Additive to baking powder to absorb dampness and keep it free-flowing.

Rice-cones

These are granular, not being so finely ground. Their main use is for dusting purposes especially for bread and rolls.

RYE

Like wheat this cereal may be milled to give varieties of flours which can be made into bread. The different products are as follows:

White Rye

Milled mainly from the centre of the grain, this product is lighter in colour than the other types. Because it has a low protein content, it lacks strength and is used for light rye breads.

Medium Rye

Flour produced from the rye grain after the bran and shell have been removed. It is darker in colour and has greater strength than white rye.

Dark Rye

This is milled from the grain after some of the finer starch cells have been removed. It has the darkest colour and is the strongest, having up to 16% of protein.

Rye Meal

We get this product by grinding the entire rye berry. It is coarse and dark and used for pumpernickel bread and rolls.

The use of these flours milled from rye to make bread is explained on page 98.

WHEAT

A wide variety of flours are milled from this cereal for use in bakery products.

White flour

This flour is milled from the endosperm of the wheat berry after the husk or *bran* and embryo *germ* are removed. Depending upon the blend of wheats from which it is milled, these flours may be weak (soft), medium or strong (hard).

Weak flour

Milled from predominantly English wheats, this flour has a low protein (gluten) content, i.e. 8%, and is ideal for many cakes, sponges and pastry. *English*, *Biscuit* and *Cake* flours are in this category, the latter being specially milled and treated.

High Ratio Cake Flour

This is specially milled to a very fine particle size and heavily bleached, a treatment that increases its moisture-absorbing properties and makes it ideal for use in high ratio cakes where the liquid content is greater than a normal cake.

High Protein Cake Flour

Like the previous flour, this has received the same treatment at the mill, but from a blend of stronger wheats. This flour is used in high ratio cakes which have to carry fruit. Some Australian flours are also soft and are suitable for cake making.

Strong Flour

Canadian and American wheats are used predominantly in the blend of wheat required for milling a strong flour because of their high protein (gluten) contents. The use of this flour is essential for goods fermented by yeast, etc.

Medium Flour

Between these two extremes we have a flour milled to provide medium strength for use in such goods as chemically aerated ones.

The strength of any flour blend can be adjusted by the flour confectioner himself, by mixing his own flour from the varieties available.

Self-Raising

This is usually a medium strength flour into which has been blended a proportion of baking powder at the rate of approx. 2% of the flour. The use of this product is not recommended because since its efficiency gradually deteriorates from the time it is made, the aerating power is always uncertain. For accurate results, plain flour and freshly prepared baking powder should be used instead.

Scone Flour

Occasionally we need to use very small quantities of baking powder in a recipe and in such circumstances it is an advantage to mix some flour and baking powder together in a fixed proportion and weigh this mixture to give the required weight of aerating agent. This is called scone flour and is compounded by mixing into 1 kg ($2\frac{1}{4}$ lb) – 65 grams ($2\frac{1}{4}$ oz) baking powder.

Hand Test for Flour Strength

The strength of the flour can be identified by squeezing it in the hand. A weak flour will cling together when the hand is opened whilst a strong flour will crumble again to a powder. This device is useful for differentiating between flours of different strengths in a bakery.

Extraction Rate Besides strength, flours vary according to their colour which in turn depends upon the extraction rate (the percentage of flour extracted). For example, if we extracted 60 parts of flour from 100 parts of wheat, the flour would have an extraction rate of 60%. White flours have extraction rates of up to 80% usually called "straight run flour". The lower the extraction rate, the whiter the flour and the higher the grade (and price). Such flours are called "top patents" or just "patents" whilst low grade flours are called "bottom patents".

Wholemeal

A true wholemeal is milled from the whole wheat berry and therefore contains the bran as well as the germ. Only stone milling can produce a true wholemeal.

Wheatmeal

Most brown flours on the market are wheatmeals made in the mill by blending various grades of bran with white flour. The added bran may be fine, medium or coarse according to the type of flour required.

Cracked Wheat

As the name suggests, this is the product from the break roller in the mill which cracks open the wheat berry. It consists of some flour together with coarse particles of semolina and bran.

Malted Meals

Wheatmeals either milled from malted wheat or made from wheatmeal with added malt flour. These are sold under various proprietary names.

Germ Meals

These are also sold under various proprietary names, e.g. *Hovis* and *Vitbe*. They are wheatmeals with added germ which is specially treated to inactivate the enzymes present. Although mainly used for bread, a number of other goods such as cake and scones may also be made from this flour.

APPROXIMATE COMPOSITION OF FLOUR USED IN BREADMAKING
(at 15% moisture content)

FLOUR	ASSUMED EXTRACTION RATE %	PROTEIN %	FAT %	CARBOHYDRATE %	CRUDEFIBRE %	DIETARY FIBRE %	ASH %	TOTAL Ca mg/100 g	Fe Mg/100 g	THIAMINE mg/100 g	NICOTINIC ACID mg/100 g	RIBOFLAVIN mg/100 g	Na mg/100 g	K mg/100 g	Mg mg/100 g	Cu mg/100 g	P mg/100 g	Cl mg/100 g	Mn mg/100 g
WHOLEMEAL	100	12.0	2.4	64.3	2.0	11.2	1.5	30	3.5	0.40	5.5	0.12	3.3	329	129	0.625	345	37	3.4
BROWN	85–90	11.8	1.6	68.5	1.09	7.87	1.37*	150†	3.6	0.42	4.2	0.06	4.0	280	110	0.35	270	45	2.5
WHITE	72	11.3	1.0	71.5	0.12	3.15	0.66*	140†	2.2‡	0.31‡	2.0‡	0.03	3.0	130	36	0.22	130	62	0.8
WHOLEMEAL (from All-British Wheat)	100	8.9	2.2	67.0	1.8	11.2	1.5	35	3.0	0.29	4.8	0.12	3.4	361	106	0.65	340	35	2.8

The addition of chalk at the rate of 235–390 mg/100 g is compulsory to all flours except wholemeal.

* Including 0.24% derived from added chalk
† Including 125 mg/100 g derived from added chalk
‡ Part of the iron, thiamine & nicotinic acid is derived from added nutrients

Ca = Calcium
Fe = Iron
Na = Sodium
K = Potassium
Cu = Copper
P = Phosphorus
Cl = Chlorine
Mn = Manganese
Mg = Magnesium

Brown flours are milled from strong wheats so that they can be made into satisfactory bread. However they may also be used for some flour confectionery items such as scones etc.

All brown flours deteriorate fairly rapidly and therefore should not be kept too long in store.

Flour Treatment

Certain substances are added by the miller to flour mainly for nutritional purposes, but sometimes to improve its baking characteristics. However, untreated flour is available.

CHOCOLATE

Couverture

This may be obtained either as plain, milk, unsweetened (bitter or block cocoa) or white. Each manufacturer will have his own particular brand of chocolate tailored to suit a particular purpose. When melted some are thick whilst others are thin and more suited to coating purposes.

Unsweetened couverture is used exclusively for flavouring sweet icings, creams, etc.

White chocolate is made from cocoa butter, sugar and milk powder. Strictly it is not couverture but it behaves in the same way.

Before couverture can be successfully used it must undergo a process known as *tempering*.

Cocoa Butter

When liquid chocolate couverture is subjected to hydraulic pressure in a filter press, the natural cocoa fat runs out and, when cold, solidifies into a hard white solid. This has several uses:

(*a*) Add to chocolate couverture to make it thinner.

(*b*) Rub into marzipan fruits to simulate the waxed nature of the skin.

(*c*) As a grease into which flower nails may be dipped prior to piping the flower in royal icing.

Cocoa Powder

Usually sold in sealed tins in two grades. The main use of this product is to flavour various types of confectionery goods.

Cake Coating

This product used to be known as baker's chocolate before this term was legally prohibited. It is made from cocoa powder, sugar and a suitable hydrogenated vegetable fat. Although the flavour and characteristic snap and gloss of true chocolate is lacking, its use finds favour because it does not require tempering and is therefore easier to use. It is not so critical to temperature changes and may be heated to 49°C (120°F). However, the viscosity changes slightly with heat, being thinner when hot than when tepid.

Other Chocolate Products

The following chocolate products are also marketed in plain and milk:

Nibs Can be used as an ingredient in cake. Also used for the ease with which it can be reduced to liquid with heat.

Buttons and Vermicelli Used for decoration.

COFFEE

Two forms of this are available for use by confectioners:
Dried Instant Available as a powder or as granules. This readily dissolves in water and can be easily measured. To preserve the flavour of natural coffee, it should be stored in a sealed container and used as soon as possible.
Extract This is usually an extract of coffee with chicory added.

If a true coffee extract is used to flavour goods in which egg whites have been used, i.e., meringues, it is advisable to incorporate a little citric or tartaric acid. Otherwise some ammonia will be released from the egg whites and turn the product a green colour.

COLOURS

Colours available for use by confectioners must be harmless and comply with the various regulations which apply in the country in which it is used. They are of two types (*a*) derived from naturally occurring matter and (*b*) coal tar products. The two most widely used natural colours are cochineal and saffron.

Cochineal and Carmine

This red colour is obtained from the dried bodies of the female insect *Coccina*. The dried powder is called carmine.

Saffron

Produced from the dried stigmas of the crocus flower. It is made into an infusion with water and used to make saffron cake which it flavours as well as imparts a yellow colour.

Coal Tar Colours

Only a very limited selection of colours in this category are permitted for use in food and are available for sale. These should always be purchased from a reputable manufacturer.

Use of Colours

Before the choice of a particular colour is made, the advice of the manufacturer should be sought. Some colours become unstable when affected by heat or acid or alkaline mediums and change colour or become bleached.

CREAM

The various types of fresh dairy cream are mainly differentiated by their butter fat contents.

Clotted – Minimum butter fat content 55%.

This cream may be used as an accompaniment to fruit and jams but does not have a function in the manufacture of flour confectionery goods.

Double – – Minimum butter fat content 48%.

When whisked, this cream may be beaten to a stiff mass which can be piped into shapes. However, for this purpose it is more economical to use the whipping cream which gives more bulk, unless a quantity of liquid, e.g. liqueur, is added for flavouring purposes.

Whipping— —Minimum butter fat content 35%.

This is the most economical cream to use for whipping purposes.

Sterilized or U.H.T.— —Minimum butter fat content 23%.

Not suitable for whipping purposes but can be used for enriching purposes, i.e. filling creams.

Single— —Minimum butter fat content 18%.

Unsuitable for whipping but can be used for enriching, filling creams etc.

Imitation

Marketed under various brand names, this product behaves in a similar way to fresh dairy whipping cream, in that it may be whipped to a stable stiff foam which can be piped. The product is made from a vegetable fat with milk solids and water. It lacks the flavour of the fresh product.

Storage

With the exception of the U.H.T. or sterilized variety, all these creams are very perishable and have a very short shelf life. They must be kept cool at approximately 4°C (40°F) and be hygienically handled because of the hazards of food poisioning.

DECORATIONS

These can be divided into (*a*) edible and (*b*) inedible artificial decorations.

Angelica

This sugar-preserved green stem is sold in lengths and is usually covered in sugar crystals. Before use the sugar needs to be washed away and the strip cut with a sharp knife into suitable sized shapes. The most common of these is the diamond which can look very attractive when used with flowers or glacé fruits.

Chocolate Shapes

Various shapes suitable for decorating torten and gâteaux are available in this medium. They are more fragile and therefore must be used with greater care.

Chocolate Vermicelli

Available in both plain and milk, this product is very useful for masking the sides of all types of decorative confectionery.

Coralettes

These are manufactured from a variety of different bases for dressing and masking the sides of gâteaux etc.

One manufacturer markets coloured nibbed almonds for this purpose calling it Almond Decor.

Crystallized Flowers

The petals of roses, violets and lilacs, as well as the flowers of mimosa, are used for this purpose. The petals and flowers are impregnated with a very bright coloured sugar syrup which crystallizes. Thus we have vivid red, violet and pink and yellow decorations which look very attractive when used.

Dragees

These small sugar balls may be obtained either coated in silver or gold.

Glacé Fruit

Cherries are the most widely used form of glacé fruit and can look very attractive with angelica.

Other glacé fruits which may be used for decorative purposes include pineapple, apricots, peaches, pears and figs. Confiture pineapple is strongly recommended for decorative purposes.

Jelly

Many decorations are made from a pectin jelly. These include:

(*a*) Orange and lemon slices (small, medium and large).

(*b*) Pineapple rings and segments.

(*c*) Various shapes such as diamonds which are coloured green and may be used as leaves in decoration.

(*d*) Imitation cherries (chellies) in yellow, red and green.

Vermicelli

Coloured icing dried in the form of short threads or very small grains are sold for decoration purposes.

Sugar

Nonpareils – these are like vermicelli made in many colours out of sugar (commonly known as hundreds and thousands).

Flowers are also available made out of royal icing, some incorporating a holder for the candles used in conjunction with birthday cakes.

Sugar Paste

A certain manufacturer has exploited the quality that this material possesses to be manipulated into decorative shapes. Colour and other detail is added by a simple printing process and the result is an attractive decoration which has a good shelf life if kept dry. Plaques for placing onto celebration cakes as well as leaves, flowers etc., are manufactured in this way.

Inedible

A wide range of inedible decorations are available to place onto various types of celebration cakes. Paper, card, plastic, plaster, metal, fabric and wax are among the materials used in their manufacture. The following list, arranged in alphabetical order, is by no means exhaustive.

Bands – silver & gold
Bells – silver & gold
Bows
Bride & groom
Candles – various colours
Cupids

Eskimos
Father Christmas
Ferns – asparagus & maidenhair
Figures – plastic
Fir trees
Flowers – all types
Heather – white, pink & mauve
Horseshoes
Inscriptions
Leaves – all types
Motifs
Ornaments – wedding cakes
Pillars – wedding cakes
Rabbits
Ribbons
Robins
Sleighs
Slippers – filled and unfilled
Snowmen
Snow scenes
Sprays – leaves & flowers
Storks – christening cakes
Vases – wedding cakes
Yulelogs

EDIFAS

This is the brand name of ethyl methyl cellulose, a substance having no food value but when made into a solution with water, can be whipped to form a stable foam from which meringues can be made by the addition of sugar. Such meringues cannot be baked in the normal way but merely dried out in a warm prover.

Edifas is also used as a stabilizing agent for sponges, meringues and japs etc.

It is used as a 5% solution in water, i.e. 50 g per kg (1 oz per pint) made up in the following way:

Boil half of the water and whisk in the ethyl methyl cellulose until clear and free from lumps. It will form a white frothy jelly. Dilute with the rest of the water. This stabilizer is not readily soluble but this method is the best way of achieving a solution. It is best left for an hour or so before use and used cold.

EGGS

All varieties of egg may be used in baking, but hen eggs are the best for confectionery purposes.

Fresh

These may be purchased in shell by number and grade depending upon the weight of the individual egg as follows:

Old Egg Grading	*New Egg Grading* *Grade*	
Large not less than $2\frac{3}{16}$ oz	1	70 – 75 grams
Standard not less than $1\frac{7}{8} - 2\frac{3}{16}$ oz	2	65 – 70 grams
Medium not less than $1\frac{5}{8} - 1\frac{7}{8}$ oz	3	60 – 65 grams
Small not less than $1\frac{1}{2} - 1\frac{5}{8}$ oz	4	55 – 60 grams
Extra small below $1\frac{1}{2}$ oz	5	50 – 55 grams
	6	45 – 50 grams
	7	40 – 45 grams

Obviously before use, the eggs need to be carefully cracked open to release the yolk and white. This should be done over a small bowl so that the egg can be examined to ensure freedom from mustiness etc., before transferring to a larger bowl. One bad egg can ruin a whole batch.

Both the whites and yolks form a skin and dry on exposure to the air. They should be covered with a damp cloth or transferred to a container which can be sealed and placed in a refrigerator until required for use.

Eggs in shell may be kept for a week or more in the refrigerator but the temperature must never drop to freezing point, otherwise the contents will freeze and the shells crack. On defrosting the egg will leak from the shell resulting in loss.

Accelerated Freeze Dried (A.F.D.)

This product is made by first freezing the egg and then subjecting it to warmth and a high vacuum so that the ice sublimes and leaves the material dry. Some loss of flavour and whipping power occurs but this is a small sacrifice to make to obtain a product with such a long shelf life. Reconstitution is effected by adding three parts by weight of water and one part by weight of A.F.D. egg.

Chilled

Preserved by storing in their shells at temperatures just above freezing point.

Double strength

These eggs are removed from their shells, whisked and condensed to half their volume to save transport costs. When used in recipes extra water must be added to compensate.

Frozen

These are marketed either in tins of approximately 3.5 kg (14 lb) or in individual sachets of smaller weights. The egg is cracked from its shell, pasteurized and frozen and maintained at $-20°C$ ($-5°F$). When the egg is required it must be slowly defrosted without the application of direct heat. It is best to stand the tin of egg in cold running water for a few hours. Alternatively it may be left in the warm bakery overnight.

Once defrosted, frozen egg must be used and never re-frozen because of the danger of food poisioning. Eggs left over should be sealed in a container and placed in the refrigerator at above freezing point. If refrigeration is not available, it may be preserved for quite a long time by mixing it with up to its own weight of sugar, due allowance being made to the recipe in which this is used.

Preserved or Pickled

If eggs are less than one week old they may be preserved for up to a year in their shell by keeping them in a 10% solution of water glass (sodium silicate)

Spray Dried

The egg is dried by atomizing in a drying chamber. This process destroys the whipping power of egg and some loss of flavour occurs. Recipes employing dried egg must rely upon chemicals for their aeration. It is reconstituted by adding 3 parts of water to one part of the dry powder.

Sugar Dried

In this process sugar is added to the egg before it is dried. This helps to preserve up to three quarters of the whipping properties of the egg. The last product available of this type in Britain had a sugar content of $\frac{1}{3}$rd and reconstitution was made by adding water in the ratio of 2 of water to 1 of powder.

Example (using imperial scale)

To make one pint (20 oz) egg–

Sugar dried egg powder	$7\frac{1}{2}$ oz	$22\frac{1}{2}$ oz
Water	15 oz	
Quantity of sugar to deduct from recipe		= $2\frac{1}{2}$ oz
Quantity of egg used		= 20 oz

Egg Albumen

These are the whites of egg and are marketed either frozen or dried.

The frozen egg whites are treated in exactly the same way as frozen egg.

Dried egg whites or albumen are available in the form of flakes, powder or crystal, the latter having the best colour. These products are reconstituted by soaking 3 parts of the dried albumen with 20 parts water. It does not go into solution readily, the time taken depending upon the product chosen, i.e. flakes up to 12 hours, crystal 3 hours, and powder 1 hour, with repeated stirring.

Specially processed albumen powders are available under brand names which may be reconstituted in the same ratios almost immediately to a solution. Although these products contain up to 99% albumen, they cannot legally be sold as dried egg whites.

Egg Yolks

These may also be obtained either frozen or dried, the latter being reconstituted by adding an equal quantity of water.

EMULSIFYING AGENTS

Although emulsifying agents have been known for over forty years, emulsifier technology has progressed rapidly since that time and there is now a great number of these products on the market, many being marketed under proprietary names with detailed directions as to use.

Lecithin

Obtained from soya bean, this is a plasticine-like paste, light brown in colour. Greater dispersion and emulsification of the fat with water can be achieved when lecithin is added. It is used in high ratio fats for this purpose.

Glyceryl Monostearate

Glyceryl Monostearate (GMS) was one of the first emulsifying agents of this type used. It extends fat by creating greater dispersion and is also used for its crumb-softening property in bread, cakes and sponges. Before adding to the mixing, it is first brought to a jelly by stirring 1 part into 5 parts of hot water, allowing it to cool before use.

Other glyceryl esters called diglycerides are available and used in the same way.

FLAVOURS

These may be produced from the natural flavouring agents of the plant, artifically made from chemicals or compounded from a mixture of both (called blended flavours).

Artificial flavours

Many flavours can be made artifically and these are relatively inexpensive. However they lack the bouquet of the true flavour and are not so stable to heat. Their use should be confined to flavouring products which are not baked.

Blended

These are compounded from a blend of natural and artificial flavours. The advantage is that the bouquet of the true flavour is partly preserved whilst the flavour is reinforced by the strength of the artificial one used.

The quality of flavours is reflected in their cost which is always a reliable guide, the best always being the most expensive.

Flavours used on their own do not always imitate the natural flavour and additions may have to be used. An example of this is in the flavour of lemon or another acid fruit where citric or tartaric acid needs to be added to simulate the true flavour.

Extracts

These are derived from the natural flavouring material by macerating the natural source in ethy alcohol. Such extracts contain the true bouquet of the material (e.g. vanilla) and because they are expensive, are reserved for use in high quality goods.

Attar of Roses and Orange Flower Water

These are produced by the distillation of the fresh flower petals with water. They have a very delicate flavour and are used in high class unbaked confectionery goods such as fondants and almond paste.

Essential Oils

Certain fruits, seeds and flowers yield an oil which contains the essential flavouring material of the plant. However for most plants it is uneconomical to extract the oil for flavouring purposes. The exceptions are in the citrus fruits in which the oil may be expressed from the caps, e.g. Lemon, Orange, Lime etc. and in spices.

These essential oils are very stable to heat and therefore their use is recommended for such products as biscuits which are baked at a high temperature.

Fruit Pastes and Concentrates

These are made from the fruit itself and whilst expensive, their use is recommended for first class products.

Lemon and orange is obtained in paste form whilst the soft fruits such as strawberries and raspberries are available as an extract.

Storage

Flavouring materials deteriorate quite rapidly and should be used as soon as possible after purchase. If left exposed to the air, oxidation occurs which will produce off flavours, therefore they should be stored in well stoppered airtight bottles. All flavours whether in liquid or powder form should be stored away from sunlight and in the cool.

(For spice flavours *see* page 30.)

FATS AND OILS (LIPIDS)

Butter

There are two types of butters – (*a*) soured and (*b*) sweet cream, the latter being available either salted or unsalted.

Soured

This is manufactured from soured cream and has a superior flavour but limited shelf life. Danish is an example of this type.

Sweet Cream

Most of the butter used by the baker and confectioner is of this type. The normal variety is slightly salted but an *unsalted* variety is obtainable and this is recommended for certain sweet unbaked and mildly flavoured goods like buttercream.

Besides flavour, the consistency of butter is also important for certain purposes. For example, in puff pastry we need a tough butter which has good layering properties. In cakemaking we require a soft butter with good creaming power.

Lard

In flour confectionery the use of lard is usually confined to savoury pastry where it imparts flavour as well as shortness. It is also a useful fat to enrich and produce a superior crumb in bread.

Margarine

Special bakery margarines are manufactured, tailored to suit different uses as follows:

Cake Margarine having good creaming properties suitable for all types of cakes.

Pastry Tougher than cake margarine, made specially for the manufacture of puff pastry.

Saltless Used for creams, icings etc.

OILS

There are three main uses for cooking oil in a bakery:

(*a*) Greasing purposes – e.g. bread tins, chelsea buns etc.

(*b*) Enriching bread – to improve the crumb and texture.

(*c*) Frying purposes – e.g. doughnuts and scones.

The most important quality of an oil is its ability to withstand being heated to high temperatures without deterioration and with the minimum absorption into the goods being fried. It is recommended that cooking oils are always purchased from a reputable manufacturer.

Some oils are marketed under their own names such as Olive, Corn, Groundnut etc., but they can seldom improve on the frying characteristics of a first class oil or fat manufactured for this purpose.

Shortenings

These are manufactured from vegetable oils to give 100% fats which have good shortening and creaming properties. There are three broad groups:

White (Compound) – A general purpose fat used for making pastry and enriching bread etc.

Yellow – Much softer in consistency, this is a high grade fat more suited to blend with margarine or butter in the making of cakes.

High Ratio – This is a high grade fat containing lecithin or super glycerinated fat to increase its emulsifying powers. Mainly used in high liquid/sugar recipes with special high ratio cake flour.

Suet

Shredded suet is available and its use is recommended for such goods as Christmas pudding and mincemeat.

Storage

All edible fats and oils commence to deteriorate from the moment they leave the manufacturer. Therefore they should never be kept too long in storage but used as soon as possible after delivery. Warmth, moisture and sunlight will hasten this deterioration and therefore all fats should be stored in a cool dry place away from sunlight.

Rancidity is partially retarded by the presence of salt so that unsalted butter and margarine have a much shorter shelf life.

FRUIT

Candied

These are fruits preserved by impregnation with a sugar syrup. Peel is treated in this way, the caps being first cured by soaking in brine, saturating in a syrup and then dried. Three citrus fruits are suitable for treatment in this way. They are orange, lemon and citron, the colour of the peel being orange, yellow and green. These peels may be purchased already cut into small dice suitable for mixing with other dried fruit for a cake, or as whole caps for the cutting to be done by the confectioner himself.

Citron peel is also available as thin slices which may be used as a decoration on Madeira cakes for example.

Glacé Cherries

Although not candied this is a sugar-preserved fruit and used with peel in mixtures of dried fruit. There are three types available:

(1) Large glacé cherries – the ones normally used for cutting up into small dice and mixing with other dried fruit.

(2) Small alpine cherries. These are useful for cherry cakes where the cherries are required to be supported by cake. They are much smaller and therefore are not so prone to sink. Being smaller these are also useful for decorative purposes.

(3) Marachino. These are normal cherries but impregnated with a marachino-flavoured syrup which enhances their flavour.

Before glacé cherries are used, the adhering syrup should be washed off and the fruit dried. Any syrup which remains in the box can be used for reducing fondant etc.

Other Fruits

Other types of fruits which may be candied include pineapple, pears, apricots, peaches, plums and figs. Also ginger, although not strictly a fruit, but a spice, is included here. Crushed pineapple and ginger are available for use in confectionery goods. All these may be used in special fruit cakes, e.g. Paradise. Some fruits, e.g. pineapple, can be easily made by the confectioner from the tinned variety.

DRIED FRUIT

Dried fruits not only refer to Currants, Sultanas, Raisins and Peel, but Figs, Dates, Apricots, Peaches, Apples, Pears and Prunes can also fit into this classification.

Currants and Sultanas

These should be of good colour and size and free from stones, stalks and dirt. To remove the stones the washed fruit should be scattered onto a metal tray a handful at a time, so that any stones present can be heard as it strikes the metal and so can be easily detected and removed. This fruit may be purchased already cleaned, but they should always receive a preliminary sorting and washing before being used. It is recommended that before use this fruit should be left in soak for 10 minutes in hot water to soften, after which it is well drained in a sieve, placed onto a clean cloth and left to dry for 12 hours or overnight. In this way the fruit will be soft and juicy and enhance the quality of the goods in which it is used. However, dried fruit should not be left too long in a damp condition before use otherwise there is a danger of mould developing.

Currants

These are the dried form of black grapes originally grown in Greece. The best quality for confectionery use (and the most expensive) is Vostizzas but very good quality currants can be obtained from Australia. When bought, the currants should be bold, fleshy, clean and devoid of any red or shrivelled berries. Containing approximately 63% sugar, currants have a good sweetening property when used in cakes and pastries.

Sultanas

Seedless yellow grapes grown in Smyrna, Persia, Afghanistan, South Australia and South Africa are used for this fruit. The picked bunches are first dipped into potash lye with hot olive oil on the surface to inhibit fermentation and to soften the skins and then dried in open sheds or in sunlight.

Sultanas should have a good bright golden colour and be fleshy and of good flavour. Some are bleached by exposure to the action of sulphur dioxide gas which acts as a preservative but these lack flavour and the traces of sulphur dioxide in the fruit can change the tone of edible colours added to any batter in which it is contained.

Australian sultanas are marketed under *crown* brands for both light-coloured and brown-coloured types. For the light-coloured type the best quality is six crown containing no dark berries. The grades go down to 1 crown containing only 50% of light-coloured berries. Again the grade is reflected in the price at which they are offered.

Raisins

There are two types. Muscatel are the large variety which have to be stoned before use, and the smaller stoneless variety. The former imparts a better flavour and is recommended for Christmas puddings etc.

Dates

If these are packed with the stones already removed, no further treatment is necessary except for dicing or mashing according to the dictates of the method of the recipe employed. Dates for manufacturing are purchased in a solid pack but desert dates are available which may be used for special goods or decorative purposes.

Figs

These may be used in the same way as dates – either diced or minced and mixed with a little water to make a paste.

FRESH FRUIT

The use of fresh fruit in flour confectionery can usually only be justified if they can be purchased at an economic price or used for goods for which no other preserved variety is available. For example, fresh strawberries are only available at certain times of the year and since they are unique in that no preserved variety can compare with the fresh product, their purchase whilst in season is desirable for various flans, torten etc. Furthermore, if a glut of fresh fruits become available, it might be a good economic proposition to freeze these supplies for future use.

Apples, Pears and Bananas

These fruits require some care in the way they are prepared otherwise they become discoloured due to oxidation when exposed to the air.

When such fruits have to be exposed to the air before use, i.e. when peeled and ready for cutting, it is best to leave them covered in salt water. Obviously the salt needs to be washed away before the fruit is actually used. Ascorbic or citric acid will also help to prevent discolouration by oxidation.

Soft Fruits

Fruits such as raspberries, strawberries, blackberries etc., are very perishable and great care must be exercised in their selection. Mould is particularly manifest in fruit which has been picked in a wet condition.

Refrigeration is essential if soft fruits are to be kept for a short period even if this is only overnight. Deterioration is soon detected by the presence of stains which appear at the bottom of the tray or punnet. They may be stored up to 3 days in shallow trays in a cool dark place or refrigerator.

Citrus Fruits

Oranges and lemons are not only purchased for their juice but also their skin which is used for the zest. They should therefore have a fresh, firm and clear peel as well as being juicy.

Other Fruits

Fruits such as gooseberries, plums etc., may be stewed and used to make all types of pies and tarts if the purchase price is satisfactory enough to take account of the economic features involved.

FROZEN FRUIT

Very few of the fruits used by confectioners will be in this category. Exceptions will be soft fruits or fruits which may be purchased because a glut has depressed the price to make it an economic proposition to purchase fresh supplies for future use. Frozen whole strawberries are available for decoration. These have been nitrogen frozen and are rather expensive.

TINNED FRUIT

Most fruit is available in tinned form and there is a wide variety of sizes as follows –

Imperial sizes		E.E.C. sizes
Size of Tin	*Approx. net weight*	
Picnic	8 oz	Equivalents
A1	10 oz	not yet
E1	14 oz	established
No. 1 Tall	1 lb	
1 lb Flat	1 lb	
A2	$1\frac{1}{4}$ lb	
A$2\frac{1}{2}$	$1\frac{3}{4}$ lb	
A10	$6\frac{3}{4}$ lb	

When selecting tins of fruit, ensure that they are not dented, punctured or blown. Rust indiçates that the tin has either been too long in store or stored in a damp place. All such tins should be discarded.

Fruit should never be left in the tin once it is opened but transferred to a clean container, preferably earthenware, and used as soon as possible. Left in the syrup,, the fruit may be safely stored for 3 days provided it is placed in a refrigerator. Prolonged storage will encourage fermentation and spoil the product. Deterioration will be more rapid in unsweetened tinned fruit.

Tinned fruit is selected according to the purpose to which it is to be put. Pineapple, for example, may be obtained in large or small slices, cubes, titbits or crush, the slices being suitable for making confiture for decorative purposes whilst the crush is ideal for various fillings.

Colour is also an important consideration particularly when tinned fruits are mixed as in flans for example. Fruits which are too pale may have their colour heightened by the use of edible added colour, e.g. green gooseberries.

Despite being tinned some fruits such as apples will brown when exposed to the air and therefore should be used immediately.

Some products are either not sweetened or require additional sweetening for the purpose for which they are required. In such products the sugar should be added as soon as possible after opening the tin to take advantage of the preserving nature of sugar which will help keep the fruit for a longer period of time.

The syrup in tinned fruit is a valuable raw material and should never be wasted. It can be used in a variety of fillings and glazes, as a medium for reducing fondant icing or to replace some of the sugar and water in fermented goods.

JELLYING AGENTS

Under this heading we include the various gums which are used either directly in the baked product, in the manufacture of prepared materials used by the baker and confectioner *or* as stabilizing agents in various confectionery products such as sponges, meringues etc. There are four main categories – animal, vegetable, marine and synthetic.

Animal Gums

Gelatine is the main animal gum we use in the bakery. This is protein in character, and is obtained from the hide trimmings and bones of animals.

It has the ability to swell in cold water and dissolve in hot water. When used in sufficient strength (i.e. 5%) a hot solution will set to a firm jelly on cooling. It is therefore used for its setting property in such goods as marshmallow, buttercream, glazes and other jelly-like preparations.

The pure edible gelatine is marketed in three forms:

(*a*) Sheet
(*b*) Flake
(*c*) Powdered or Crystal.

A good grade of gelatine should be a pale golden yellow, free from odour or flavour and capable of making a perfectly clear and bright jelly.

Although gelatine will readily dissolve in hot water, prolonged boiling should be avoided as this will reduce the strength of the jelly formed.

Vegetable Gums

Arabic (Acacia) is the exudation from a number of species of Acacia tree. The best quality is colourless or whitish and is sold as a dry powder which needs reconstituting with water to form a gummy solution. The main use of this gum is for glazing almond products, e.g. parisian rout biscuits.

Tragacanth This is an exudation from a species of shrub. It is marketed as flakes or as a finely ground white powder which, when water is added, will not dissolve but swell to form a mucilage which is much stronger than gum arabic. Its main use in confectionery is to make gum paste (pastillage), stiffen almond paste for modelling purposes and as a stabilizer for various emulsions.

Gums Ghatti and Karaya These are also to be found in various proprietary foods and are often used as substitutes for arabic and tragacanth.

Carob (or locust bean) Extracted from the seeds of a tree of the bean family, this can be used as a substitute for gum tragacanth for a number of purposes and has high emulsifying powers.
Quince This is obtained from the seed of a shrub, the flowers or fruits of which are very similar to apples or pears. The gum which is extracted by agitating 2 parts of seed with 100 parts water for half an hour, is approximately 10 times as effective as gelatine as a stabilizer for ice cream.

Marine Gums

These originate from algae or seaweed which is grown around the shores of Great Britain, Japan and the U.S.A.
Agar Agar This is produced from a red algae and is obtained either as strips or a powder forming a jelly when boiled with water. It is used extensively for making all types of products, e.g. jellies, sweets and marshmallow. It possesses eight times the gelling power of gelatine, a good jelly being obtained with a $\frac{1}{2}\%$ solution. Prolonged storage must be avoided, however, as ageing reduces its swelling power.
Irish Moss (Carragheen) Derived from seaweed found around the British Isles and the Atlantic coast, the dried extract of this gum is available for sale commercially and is mainly used as a stabilizer in emulsified products, i.e. cream.
Alginate Another gum derived from seaweed. It is converted into sodium alginate and this commercial product is marketed under the name of *Manucol*. Its main use is as a stabilizer for various emulsions like cream.

Synthetic

These have a basis of cellulose derived from wood pulp. They possess very high emulsification and stablizing properties and are used in many proprietary foodstuffs.

Pectin

Although not strictly a gum this nevertheless forms a jelly when it is boiled with sugar in the presence of acid. It is the natural jellying agent found in fruits and vegetables which makes it possible to form jams and jellies from these materials. For a perfect pectin jelly to be made, it not only requires a sugar concentration of approximately 65% but sufficient acid to bring the jelly to a pH of 2·0 to 3·5. In fruits deficient in acid, this has to be made up by adding citric or some other edible organic acid. Deficiency of pectin can also be made good by adding this substance.

Two forms of commercial pectin are available:

(1) Liquid – usually produced from apples.

(2) Powdered – obtained from citrus fruits. This is the most concentrated form available.

Normally a concentration of 1% pectin is adequate to produce a firm jelly if sufficient acid and sugar are present.

Pectin Jelly

A product known as quick set jelly is marketed for confectioners to use as a glaze for fruit flans. It consists of a heavy sugar syrup in which the pectin is in solution. This remains as a liquid until a measured quantity of the complementary acid solution (usually citric) is stirred in, when it rapidly sets to a transparent jelly.

Starch

When starch is heated in the presence of water it swells and eventually forms a gelatinous paste. This property is used for the making of custards and various glazes in the bakery. Two products used extensively for this purpose are Cornflour and Arrowroot:

Cornflour or Cornstarch Produced from maize, this product is used at the rate of 5% for making such goods as custard.
Arrowroot This is a white powder obtained from the tubers of the Maranta plant. It is used mainly to thicken fruit juice with which it is heated to form a glaze for fruit flans etc. For this purpose it is superior to cornflour since it sets to a more transparent jelly.

Modified and Waxy Starches

These products can also be used with advantage in pie fillings which are likely to be frozen since they have great stability against freeze/thaw which for ordinary starches would result in weeping.

MALT

Extract

For use by the baker and confectioner, several types of this product are available depending on their diastatic or lintner value.

High diastatic	Lintner value approx. 100°
Medium diastatic	Lintner value approx. 65°
Low diastatic	Lintner value approx. 30°
Non diastatic	Lintner value nil

The extract is in the form of a thick viscous syrup the low diastatic variety being pale in colour whilst the high diastatic is much darker.

Because of the high maltose sugar it contains, malt extract has a long shelf life provided it is sealed away from insects.

Dried Extract

This is the viscous extract dried to a powder in a vacuum oven which reduces its volume as it is concentrated.

The product is very much easier than the viscous extract to mix into batters etc., because it can be sieved with the dry ingredients.

Its one disadvantage is that being extremely hygroscopic it will rapidly absorb moisture from the air to revert to the viscous form, unless kept in an airtight container in a dry store.

Malt Flour

When the malted grain is milled into flour we get this product. It is obviously not so concentrated as the other products, but is more convenient to use to blend with other flours either as an improver or to manufacture malted proprietary flours.

MILK

Fresh

The types of fresh cows' milk available in Britain are Pasteurized, Sterilized, Ultra Heat Treated and Untreated. Some may be designated Channel Island, South Devon etc. to show that it comes exclusively from this breed of cow.

Homogenized

This is milk which has been forced under pressure through very fine holes to reduce the size of the fat particles. The result is a perfect homogenous mixture in which the cream cannot separate out.

Pasteurized

Approximately 95% of all milk sold by retail in Britain is pasteurized. In this process the milk is generally heated to not less than 72°C (161°F) for at least 15 seconds and rapidly cooled. This preserves the life of the milk by arresting the lactic acid bacteria responsible for the souring of this product.

Sterilized

This is homogenized milk which is first bottled and then subjected to at least 100°C (212°F) to ensure a long shelf life. Although it keeps for a very long time, this treatment coagulates some of the milk protein and confers a cooked taste to the milk.

Ultra Heat Treated

U.H.T. milk is heated for a second or two to at least 132°C (270°F) and then filled immediately and aseptically into containers, giving it an indefinite life.

Untreated

Milk which has not received any type of heat treatment.

MILK POWDERS

Two methods may be used to reduce milk to powder. These are:

Roller Dried The milk is spread onto heated rollers from where it is removed by a scraper and ground to powder. This gives a coarser granular powder of poorer colour which is not so soluble in water.

Spray Dried In this process the milk is atomized and sprayed into a heated chamber under vacuum, where it is immediately reduced to powder. This produces a much superior product which readily dissolves in water.

There are three types of Milk Powder depending upon the fat content:

Full Cream

This reconstitutes into a liquid almost identical to fresh milk by using one part to eight of water. Because of its high fat content it has a tendency to go rancid on prolonged storage and it should therefore only be purchased in small quantities and kept in a cool store.

Half Cream

Similar to full cream, this product has half its fat content removed before drying, being reconstituted at the rate of one part to nine parts of water. Its shelf life however is still very limited.

Separated or Skim Milk Powder

Removing the cream from the top of the milk by skimming prior to making it into powder, we get this product. It is reconstituted at the rate of one part to ten parts of water. Because it can be kept for much longer periods of time without deterioration it is much more popular with food manufacturers. The cream which has been removed may be put back in the form of butter along with the skimmed milk powder in certain goods, e.g. milk bread.

Buttermilk

This is a by-product obtained in the churning of butter from ripened cream. It contains approximately 90% water and 10% milk solids of which approximately 1% is lactic acid. However it may also be obtained in the dry form. This product may be used in aerated goods where it can replace some of the acid material of the baking powder.

Whey Powder

The whey is the by-product of the cheese manufacturer and this may be dried and used in various goods. It is extremely rich in lactose sugar and will store almost indefinitely if kept dry.

Modified Powders

Several proprietary powders are marketed under brand names. They are made from separated milk powder and vegetable fat which gives them a longer storage life and they can be more easily reconstituted.

CONDENSED AND EVAPORATED MILK

The milk is concentrated by heat which not only reduces bulk but also gives the product much longer keeping qualities. Both whole and skim milk may be used for this purpose but the ratio of fat and fat-free solids of the whole milk must first be standardized at a ratio of 1:2·44.

Condensed

This is made by first heat-treating the milk, adding sugar to a concentration of 42–45%, condensing and then packing into sterile containers. Condensed milk is usually used by confectioners for enriching certain goods, e.g. fondant.

Evaporated

Sugar is omitted and the milk is homogenized before condensing, otherwise the manufacturing process is similar. Evaporated milk may be reconstituted by adding an equal volume of water. It is more usual, however, to use it straight from the tin to enrich various confectionery goods, e.g. creams.

These two products only have an unlimited life whilst they remain sealed in the tin. Once opened, they will deteriorate in the same way as milk although not so rapidly, the condensed enjoying a longer shelf life than the evaporated, but the sugar tends to crystallize out so that the product should be stirred before use.

All these milk products can be used to replace the natural milk used in goods but allowances have to be made for either their deficiencies of fat and/or water and any additives such as sugar and the recipe altered accordingly. The following chart showing the composition of the various types of milk products may be useful for this purpose.

COMPOSITION OF THE VARIOUS TYPES OF MILK PRODUCTS

	Water	Butter Fat	Protein	Milk Sugar (Lactose)	Mineral	Cane Sugar
Type	%	%	%	%	%	%
Whole fresh	88·00	3·50	3·25	4·50	·75	Nil
Whole dried	1·50	27·50	27·00	38·00	6·00	Nil
Whole evaporated	72·00	8·00	7·25	10·50	1·75	Nil
Whole sweetened condensed	31·00	8·00	7·75	10·50	1·75	41
Skim liquid	91·00	trace	3·50	4·75	0·75	Nil
Skim dried	2·50	1·50	36·00	51·50	8·00	Nil
Skim evaporated	72·00	trace	11·00	14·50	2·50	Nil
Skim sweetened condensed	29·00	trace	11·00	14·50	2·50	43

MOULD INHIBITORS

The practice of wrapping bread and confectionery in film has increased the incidence of mould (*see* page 59). This can be reduced by the use of a number of substances which come under this category. The amount which may be used is controlled by legislation and is based upon an amount not exceeding a number of parts per million calculated on the weight of the finished, baked and cooled product.

Substances recommended for this purpose are as follows:

Calcium propionate	not exceeding	1,250 p.p.m.
Calcium sorbate	not exceeding	1,170 p.p.m.
Potassium sorbate	not exceeding	1,340 p.p.m.
Sodium propionate	not exceeding	1,290 p.p.m.
Sodium sorbate	not exceeding	1,190 p.p.m.
Sorbic acid	not exceeding	1,000 p.p.m.

The latter is recommended by some to give the best results.

NUTS

General

All nuts have a very limited storage life because of their high fat content which very rapidly goes rancid. Moths are also a hazard and if the eggs are already laid in the nuts prior to delivery, it is not long before they hatch and the grubs start eating the nuts.

They should always be purchased in small quantities and used as quickly as possible. If they have to be stored for any length of time they must have a *cool* store and be kept in air-tight tins.

ALMONDS

These are the most important nuts used in confectionery. There are two types – bitter and sweet.

Bitter almonds are mainly used for producing the essential oil which is used for flavouring purposes. A small amount – up to 5% – is often added to various almond mixtures such as marzipan to enhance the flavour.

Almonds both bitter and sweet are grown in Southern Europe, parts of Asia and California. Most of the supplies coming into Britain are grown in Spain and Portugal; Jordan almonds are regarded as the best.

Sweet almonds are generally used in confectionery and this can be obtained in many different forms as follows.

Whole unblanched

As a decorative medium these look very attractive on gâteaux, torten or fancies. If used in this way they should be selected for uniformity of size and shape.

Split

The nuts are blanched by placing into boiling water, removing the brown skins and splitting in half by machinery. These are used as decoration on fancies, gâteaux etc. and on rich fruit cakes, e.g. Dundee. They may be placed onto the cake either with the round or flat side showing.

Flaked

These are sliced very thinly but still retain the shape of the almond. They are also used for decoration for tops of cakes and for masking gâteau and torten when they may be roasted to a golden brown prior to use.

Strip

For these the split almonds are cut to form strips which are used for decoration in the same way as flaked.

Nibbed

The whole blanched almonds are chopped up into small pieces known as nib almonds. These are available as small or large nibs. Besides being an ideal decorative medium used either in their natural state or roasted, they are also used as an ingredient in cakes. They may also be coloured with edible liquid colour and dried and used for decoration. One well known firm markets this product in various colours as almond decor.

Ground

In this product the almond is ground to form a coarse powder without releasing the oil. Cheaper varieties of ground almonds are available which are made from inferior nuts and sometimes adulterated with apricot and peach kernels. A very small amount of bitter almonds is sometimes added to enhance the flavour.

ALMOND PRODUCTS

Marzipan

Legally, this also refers to almond paste and almond icing but if sold as such it must contain not less than 25% almond content (equivalent to 23·5% almond solids) and no other nut product. Not less than 75% of the remainder shall be solid carbohydrate sweetening matter. Several products sold either as various types of almond paste or marzipan are sold subject to compliance with this legal definition.

Raw Marzipan

This is a grey paste made by grinding to a paste $\frac{2}{3}$rd almonds which have been previously steeped in water and $\frac{1}{3}$rd sugar. Sufficient moisture has been absorbed by the almonds to form a paste which is then cooked to a temperature of 104°C (220°F) until it has reached the correct consistency. This paste is very smooth and may be used in several ways as follows:

(1) Added to cake for the following:
   (*a*) Flavouring.
   (*b*) Assisting to keep the cake moist.
   (*c*) Enriching purposes.
(2) Made into a modelling almond paste by the addition of icing sugar.
(3) Made into various macaroon and almond goods.

When purchasing marzipan or almond paste it is essential to know the percentage of almond it contains, not only for the purpose of comparing prices of individual products, but also to know how they should be used to produce other almond goods or the quantity of sugar which should be added.

Once exposed to the air the surface of almond paste and marzipan rapidly dries to form a crust. Care should therefore be exercised on the way it is removed from the pack. It should be first emptied out of its container, the wrapping removed and the approximate amount required cut cleanly from the block. It should always be wrapped again, preferably in polythene to prevent skinning.

Macaroon Paste

Another almond product sold to confectioners is macaroon paste. This is actually a stiff macaroon biscuit mixture containing almonds and sugar in the ratio of 1:2. All that is necessary to add to this mixture to make it suitable for various types of almond products is egg or egg whites.

Nougat (or Praline Paste)

This paste is made by adding roasted almonds and/or hazelnuts to a boiled sugar solution and the whole mixing ground to a paste by passing through granite rollers.

It is sold as a paste for flavouring purposes principally in creams.

Nut and Kernel Paste

This term indicates pastes in which other varieties of nuts may be used, e.g. apricot and peach kernels and coconut.

OTHER NUTS

Brazil

This high quality dessert nut from South America is usually used either as a decoration or made into a paste with sugar to form chocolate centres.

Cashew

The shape of this nut which is in the form of a kidney makes it very useful for decorative purposes. It is usually used roasted – a process which improves its eating quality.

When ground, it makes a very acceptable substitute for ground almonds.

Chestnuts

The form in which these are used is the marron glacé. This is made by first cooking the chestnut in boiling water and then sugar solutions of gradually increasing concentrations until the nut is impregnated with sugar. A paste made by grinding the marron glacé may be used for flavouring purposes.

Coconut

Although this nut produces a very high grade oil which is used for making margarine, cooking oils and cosmetics, it is the use of the fleshy "meat" which is of more interest to confectioners. This is dried (desiccated) and then processed into various types of products as follows:

(*a*) Fine desiccated.
(*b*) Medium desiccated.
(*c*) Coarse desiccated.
(*d*) Thread, shred or strip.
(*e*) Flour.

Products *a–d* may be used as ingredients in such goods as coconut meringues, but its main use is for decoration; for this purpose it may be used plain, roasted or coloured with edible colour and dried.

Coconut flour is used either as an ingredient to flavour various confectionery goods or made into a paste with sugar and used in the same way as almond paste.

Ground Nuts (Arachis, monkey or peanut)

The oil from the groundnut is valuable for the production of margarine and cooking oils, but the nut itself can only be used as a cheap substitute for almonds. It may be used in cheap almond substitute pastes or as flaked for decorative purposes.

Hazelnuts (Filberts and Barcelonas)

These are all derived from the same family differing mainly in size and shape, imported from Italy, Spain and Turkey. They may be used whole, either blanched or unblanched, or roasted for decorative purposes. The unblanched hazelnuts may be ground to a paste either plain or roasted and used to enhance the flavour of various confectionery products. Hazelnuts may be used to replace almonds in all recipes in which they are used. Ground hazelnuts are also available.

Pecan

This nut which is a native of North America resembles a walnut but is much smaller. It may be used in the same way as walnuts either as an ingredient in cakes and bread or as a decorative medium. In this latter role it has greater advantages over the walnut because of its smaller size.

Pinenuts (Pignolia)

Comprising the seed kernels of various species of pine trees, these small nuts are very rich in oil and have an attractive resinous flavour. They may be used in the decoration of flour confectionery.

Pistachio

Called green almonds, these nuts are cultivated in the Mediterranean region. They are green in colour and about $2\frac{1}{2}$ cm (1 in) in length. After blanching to remove the skin, they may either be used whole or nibbed for decoration purposes in high class confectionery. Because of their high cost their use as a decorative medium has largely been replaced in Britain by the use of strip almonds coloured green.

Walnuts

These nuts are cultivated in the central and southern regions of Europe. When added to bread and cakes they impart a very distinctive flavour which some customers find attractive. They may be purchased and used, crushed or broken as an ingredient in cakes and bread or whole for decorative purposes. If used for this latter purpose, selection has to be made to ensure uniformity of size.

Replacement of Almonds by other nuts

Most almond goods can be made substituting the almonds with another variety of nut with slight adjustment to the recipe to allow for the moisture and fat contents.

The following table shows the approximate composition and energy value of the various nuts mentioned.

	Water	Protein	Fat	Carbo Hydrates	Cellulose	Mineral Salts	Energy Value kJ/100 g
Almond	4·8	21·0	54·9	14·3	3·0	2·0	2751
Brazil	4·7	17·4	65·0	5·7	3·9	3·3	2935
Cashew	5·2	12·1	44·3	17·0	1·3	2·6	2230
Chestnuts	43·4	6·4	6·0	41·3	1·5	1·4	1054
Coconut	13·0	6·6	56·2	13·7	8·9	1·6	2545
Coconut dessicated	3·5	6·3	57·4	31·5		1·3	2780
Ground Nut	7·4	29·8	43·5	14·7	2·4	2·2	2465
Hazlenut	5·7	12·9	64·0	5·2	12·0	—	2809
Pecan	3·4	12·1	70·7	8·5	3·7	1·6	3116
Pine Kernel	6·2	33·9	48·2	6·5	1·4	3·8	2688
Pistachio	4·2	22·6	54·5	15·6		3·1	2717
Walnut	3·4	18·2	60·7	13·7	2·3	1·7	2919

The above table is adopted from the book *Nuts* by F. N. Howes, published by Faber & Faber. (kJ = kilojoules).

PAPER AND WRAPPING MATERIALS

Aluminium Foil

For many years the Baking Industry has been using aluminium foil containers in which all types of goods from pastry to bread can be baked. They have the advantage that the use of tins is eliminated with the need for cleaning, hence they are more hygienic. Also with fragile goods such as custard tarts, since the container is sold with the product, breakages are reduced to the minimum because the goods do not have to be removed from the tin. Their one disadvantage is that the baking temperature or time has to be increased slightly to overcome the heat reflective power of the foil.

Ovenable paperboard containers

A revolutionary alternative to aluminium foil is now on the market. This is a paperboard which is capable of withstanding deep freeze and cooking temperatures up to 204°C (400°F). The paperboard is preformed into a container of any shape and size to suit the goods in which they are baked. They are coated with a polyester coating which resists heat and also provides a barrier against grease and liquids, so that the goods have a good clean release from the container. Another bonus is that the container can be printed with heat resistant inks making them not only a functional container during the production and storage of the food, but also providing an attractive display and serving dish. They also have the advantage over aluminium that they do not dent, being more rigid, and can be easily stacked in store.

With the price of paper remaining stable and the price of aluminium rising, this latest aid to bakery production would seem to have a profitable future.

Cap

This is a thin sulphite paper for general purpose use. It is useful in manufacture for weighing recipe materials upon.

Cartridge

A thicker and better quality paper than cap, this may be used to protect rich cakes against the heat of baking.

Corrugated

Useful for heat protection when baking rich cakes since it also incorporates air in its folds.

Greaseproof

This is a very versatile paper used for lining cake tins, making icing bags and upon which ingredients may be weighed. There is usually more than one grade available, the strongest and thickest being the best (and most expensive).

Silicone

Similar to greaseproof, this paper is impregnated with silicone which has superb release properties for goods which are baked upon it. Products which are normally difficult to remove from paper such as those containing a high sugar content may be baked upon this material with advantage. At least two grades are available – thick and thin. By wiping over with a damp cloth to remove surplus sugar etc. from a previous baking, this type of paper may be used for a number of bakes before being discarded. The thick paper is able to stand up to more bakes than the thin, but of course is more expensive. Heat penetration is reduced by the use of this type of paper and the bottom heat of the oven may have to be increased to compensate.

Wafer

Known as 'rice paper' since it is made from this cereal, this may be obtained either in large sheets to fit a normal size baking tray, or in smaller sized sheets. Wafer paper is edible and is used in many types of macaroon goods.

As it dissolves in water it is essential to keep this paper dry in which condition it will keep for a very long time. Being edible it must be kept sealed away from insects, rodents and birds.

Waxed

This is paper which has either been impregnated with or coated with wax on both or only on one side to give it protective and heatsealing qualities. Waxed tissue is useful for the execution of delicate run-out sugar off pieces in cake decoration.

Transparent Film

Materials in this category in general use are Acetate film, Polythene and Cellulose film: but the most widely used of these is the latter.

Cellulose Film

Apart from the transparency of this film its most obvious advantage for bakery goods is the hygienic protection it offers against dust and germs. The other advantage is its protection against staling which is offered by choosing the grade of film appropriate to the type of goods which are required to be wrapped. The uses of four basic grades of film are discussed here:

Plain Transparent (P.T.) This is a breathing, uncoated non-heatsealing film mainly used for goods with a short shelf life, e.g. cakes coated with cream, fruit, sugar etc. where protection against dust, grease and handling is the main concern. Sealing has to be done by gum, tape of staple.

Greater Permeability Heatsealing Anchored Transparent (GSAT) Being lacquered on both sides this is a breathing film suitable for wrapping the same kind of cakes as the P.T. but can be heatsealed.
Lacquered Heatsealing Anchored Transparent (LSAT) This is a semi-moistureproof film coated on both sides. The moisture vapour transmission of this grade is less than that of the GSAT grade and is used for wrapping such goods as sponge sandwiches with a high moisture content, but with a sugar content insufficiently high to act as a protection against mould. It is important in such goods to retain as much moisture as possible to delay staling but prevent the build up of a high humidity inside the film which would encourage mould.
Moistureproof Heatsealing Anchored Transparent (HSAT) The grade of film known as MSAT gives full protection against the transfer of moisture either from the cake or directed to it from the atmosphere. It is used mainly for goods with a low moisture content like biscuits or high sugar and/or fruit cakes whose ingredients inhibit mould growth and where it is desirable to maintain their moist cake texture.

Colour printing incorporating the name of the firm and product in attractive designs can also be incorporated in these wrapping materials.

SALT (SODIUM CHLORIDE)

The type of salt usually used by bakers and confectioners is vacuum salt which has a fine crystal structure and will readily dissolve. It is important to store salt in a dry store otherwise it will become compacted into a hard mass. Salt aids corrosion of most metals so it is recommended that it is stored in plastic containers.

SOYA

Soya flour is a very nutritious product milled from the soya bean, which is a stable article of food in the Far East. Two types are available – processed and unprocessed.

Processed soya is heat treated to render its enzymes inactive, eliminate the bean flavour and permit a longer storage. It may be used in confectionery to reduce costs.

Unprocessed soya may be used in bread making (*see* page 46) where its active enzymes can speed up fermentation. Besides contributing to a more thorough ripening of the dough its use will enable more water to be added. It is claimed that soya will improve the softness of the crumb, delay staling, besides increasing yield.

Defatted Soya

Full fat soya has 20% oil and 40% protein. By removing the oil we can get a product with only 1% oil and 59% protein. This fat free soya resembles egg and is capable of producing whipped products such as marshmallow.

Storage

The unprocessed soya flour rapidly deteriorates and develops rancidity. Therefore only small quantities should be purchased at a time and kept in cool conditions.

Defatted soya has a longer storage life than processed soya.

SPICES

Spices are aromatic or pungent vegetable substances used to flavour food. They are obtained from various parts of different plants as follows:

Barks	–	e.g. Cinnamon, Cassia
Buds	–	e.g. Cloves
Flowers	–	e.g. Roses & Orange flower
Fruits	–	e.g. Aniseed, Peppers
Leaves	–	e.g. Sage, Thyme
Roots	–	e.g. Ginger
Seeds	–	e.g. Nutmeg
Stems	–	e.g. Angelica

Spices owe their aroma and pungent flavours to the presence of various essential oils and glucosides. These can be extracted by alcohol and other solvents to produce essential oils and flavours (*see* pages 13 and 14).

Allspice (Jamaica pepper)

This spice gets its name from the fact that its flavour resembles a combination of cloves, nutmegs and cinnamon. It is made from the ground, dried, unripe berries of an evergreen which grows in Jamaica.

Aniseed

The seed-like fruits of a small plant indigenous to the Eastern Mediterranean region, this spice is mainly used as a dressing on speciality bread and rolls.

Caraway

These seeds are from a carrot-like plant which grows throughout North and Central Europe and Asia. They are $\frac{1}{4}$ in long, curved and almost black in colour. They are used in seed cakes and speciality bread and rolls.

Cardamom

Consists of the whole or ground seeds of a herbaceous perennial of the ginger family and is a native to Southern India where it is produced.

Cinnamon

Obtained from the bark of two year old shoots of a species of laurel tree grown in Ceylon. The bark is carefully scraped and formed into long cylindrical quills, pale yellow brown in colour. The ground spice is used either on its own or as part of mixed spice.

Cloves

Consisting of the dried flower buds of an evergreen, this is a native of the East Indies. The pungent flavour is due to the phenol called eugenol which can be extracted and made into synthetic vanilla essence. The dried bud can be used on its own to flavour apple in tarts etc. or ground and used in spice blends.

Coriander

This is the fruit of an annual which is native to Southern Europe. The ground spice is used in various spice blends.

Ginger

The rhizomes or underground stems of a reed-like plant produces this spice. It is cultivated in India, Japan, Nigeria, Australia and Jamaica from where the finest is grown. It may be preserved in sugar and used as a fruit or the dried rhizome may be ground to a powder and used on its own as in ginger goods or in spice blends.

Mace and Nutmeg

Both of these come from the same tree, the mace being the dried seed coat and the nutmeg being the seed. The tree which grows in the East and West Indies produces a peach like fruit which, when ripe, splits open to reveal the orange-coloured seed coat. This is removed and dried in the sun where it changes to a buff colour and is known as Mace. The seed is also removed and dried, the hard coat or testa being removed to leave the nutmeg. This is then steeped in sea water and finally smoked. Nutmegs are graded for size as 40, 65, 80 etc. being the approximate number per lb. Both mace and nutmeg have a similar flavour. The ground spice may be used on its own or in spice blends.

Maw Seeds

These are bluish grey in colour and are used sprinkled over speciality bread and rolls.

Pepper

These are the fruits of a climbing vine grown in the Far East. They are gathered when red and then dried when they turn black and wrinkled. For white pepper the black skin is removed after soaking in lyre. White pepper is less pungent than black.

Poppy Seeds

The flowers from where these seeds are obtained grow wild in great abundance throughout Europe. Whole poppy seeds are used sprinkled on speciality bread, rolls, cakes and pies.

Sesame Seeds

Another native of the Far East, this seed is also used to sprinkle onto speciality bread, rolls and cakes.

Mixed Spice

This varies with the manufacturer. Two typical recipes are as follows:

	Recipe No. 1	*Recipe No. 2*
Rice Flour	25%	—
Cinnamon	28%	32%
Caraway	25%	—
Coriander	3%	32%
Ginger	3%	16%
Mace	11%	—
Nutmeg	5%	16%
Pepper	—	4%
	100%	100%

Storage

The chemicals which are responsible for the pungent flavour of spices are very volatile and reactive, especially to oxygen. It is very important therefore that spices should be kept sealed in airtight containers, purchased in small quantities at a time and used as quickly as possible. If left too long before use, the aromatic qualities will be lost. It is also important to carefully choose the place where spices are stored. Aroma and flavour can easily be transferred and modified. Spices therefore should not be stored close to materials having delicate or bland flavours like cream to which it could transfer its flavour. Also it should not come into proximity with other strong odours, e.g. onions.

STABILIZING AGENTS

These are substances which when added to unstable mixtures such as sponges etc. tend to stabilize the foam and prevent the air which has been incorporated in their manufacture from escaping, causing a breakdown.

Most of them are gums (*see* under Jellying Agents page 19) such as gum arabic or ethyl methyl cellulose (Edifas page 11) which is the main agent in at least one proprietary product.

Their use is well recommended in those products which rely upon unstable foams such as egg and sugar with fat or flour/ground almonds, e.g. japonese.

SUGAR AND SUGAR PRODUCTS

By the term sugar we mean *sucrose*, the sugar derived from sugar cane or sugar beet and mainly used for the making of cakes etc. Some of the sugar products include other sugars principally laevulose (fructose) and dextrose (glucose), but are included because they are in general use by flour confectioners.

Castor Sugar

Probably the most important grade of sugar as far as the confectioner is concerned because it is used for most types of cakes and pastries. There are three grades available – Fine, Medium and Coarse, although the latter is almost indistinguishable from the fine granulated. The fine castor is more in demand because it more readily aerates a mixture when beaten with fat and is also more easily dissolved.

Cube

This is available as small and large cubes. Because finely divided sugar crystals can easily collect dust, cube sugar is reputed to be the purest form obtainable. Therefore the large cubes are sometimes recommended for sugar boiling in which we require a sugar syrup free of any scum.

Demerara

Comprising of large light brown crystals, this partially refined sugar is often used as a decorating and dressing medium for fermented goods, biscuits etc.

It is rarely used as an ingredient because of the difficulty in getting it to dissolve.

Fondant

This is a white icing which is marketed in solid packs. Before use it requires to be reduced to the correct consistency with either stock syrup or some other liquid and heat. Prolonged storage will make this icing go hard as its moisture evaporates and makes it very difficult to use successfully. It is recommended that on delivery, the pack of fondant should be transferred to a container which can be made airtight. A little water poured on top ensures that no hardening of the surface of the fondant takes place in store.

Dry Fondant

Fondant is now available as a dry powder similar to icing sugar and is reconstituted by adding water or fruit juice. The use of the latter makes an icing which is of a much superior flavour.

Confectioners' Glucose

Confectioners' glucose must not be confused with the pure dextrose which has this name. It is a product manufactured from starch consisting of a complicated mixture of different sugars and the gum dextrin in the form of a thick, viscous, clear, transparent syrup. The main use of confectioners' glucose is in sugar boiling.

There are several grades available according to their dextrose equivalent (DE) e.g. 34, 42, 55 and 63. The types having low DE figures have a low sugar content with a correspondingly high percentage of dextrin gum and are the types which should be used for sugar boiling, since the dextrin has an inhibiting effect on the premature crystallization of sugar.

Conversely, glucoses having high DE figures have a low dextrin content and high percentage of glucose sugar. Since this dextrose is very hygroscopic the use of this type of product in cakes is recommended to keep them moist.

Dried glucoses (also maltose) are now available for use in baked goods and are likely to supersede the viscous types. They can also be made with a predetermined composition of various sugars to suit different recipe formulations. The great advantage of these is the ease with which they may be incorporated into a cake – sometimes a difficulty experienced with the viscous glucose syrup. Because of its hygroscopic property the dried type needs careful storage to keep it a free-flowing powder.

Confectioners' glucose has an indefinite life, although the viscous types may become cloudy on prolonged storage.

Golden Syrup

This is a perfectly clear and transparent syrup, pale amber in colour. It should never be kept too long in storage before use since crystallization can occur and the syrup may become discoloured.

Granulated

Usually this is the cheapest form of sugar available and therefore is used whenever possible. Because of its large crystal size, it does not readily dissolve when incorporated into a cake but must be first dissolved in water or milk. Its use is therefore confined to cheap cakes, fermented goods, syrups etc. There are three grades – fine, medium and coarse, the latter being less expensive.

Honey

Obtained either as a clear, golden-coloured, thin syrup or as an opaque crystalline mass. For use in flour confectionery the thin clear syrup is recommended since it readily mixes in with other materials. The thick honey is the crystallized variety and more suited as a preserve.

Whilst honey will keep indefinitely, the thin variety will gradually crystallize on prolonged storage. When purchased it should be examined for the presence of crystals.

Icing

This is merely crystalline sugar crushed to a powder. It should be pure white in colour, free from any grittiness and free-flowing. Sometimes a small quantity of calcium chloride is added to absorb any moisture and improve its free-flowing properties. Usually a manufacturer will offer two or three grades depending upon the fineness of the milling. The coarse grade approaches pulverized sugar (see later). The grades are reflected in the price, the best and most expensive being the finest grade. For royal icing the superfine grade is recommended whilst for cakes, the poorer cheaper qualities may be used.

On storage, icing becomes compacted and will eventually become a hard mass, especially if any dampness is present. Ideally it should be purchased in small quantities at a time and stored for no longer than necessary.

Invert

A thin, clear, transparent syrup (a mixture of glucose and fructose), this product is useful for its hygroscopic property and is used in cakes. Because honey is really invert sugar made from the nectar collected by the bees, this product tastes very similar.

Nib

This is sugar crystals in clusters to form granules about 3 mm ($\frac{1}{8}$ in) diameter. Its main use is in decoration and dressing of certain baked goods, e.g. bath buns.

Preserved

Consisting of large broken pieces, this product is used mainly for sugar boiling and jam manufacture instead of the cube sugar which is more expensive.

Pulverized

This is really a poor grade of icing sugar having a coarse, gritty texture and grey colour. It is useful in shortbread and biscuits where it can more readily dissolve in the small amount of liquid available.

Raw

Several grades of raw sugar are available. Some are very light brown in colour whilst others are very dark. They have a fine crystal structure and are ideal for use in rich celebration cake such as Wedding and Christmas where they not only contribute colour to the crumb but also flavour. The darker the sugar, the more pronounced this flavour will be.

Raw sugar may be purchased under a number of different names – Thirds, Fourths, Pieces, Moist Barbados, Muscovado etc.

Since the sugar crystals are coated with black treacle, this type of sugar will form into a hard lump if left too long in store.

Treacle

This is very dark in colour and has a pronounced attractive flavour. Cane molasses are very similar to black treacle in both appearance and taste. Both can be used in flour confectionery for flavour and colour and for their hygroscopic properties.

Storage of Sugar

All sugars require careful storage conditions. They should particularly be kept free from any dampness which will cause the crystals to coalesce and form a solid compacted mass. Insects, rodents, birds and animals all love to consume sugar. Therefore on delivery, it should be put into lidded containers to keep it from contamination.

Standards for Sugar Grain Size

The choice of a particular grade of sugar nearly always depends upon the size of the grain and this tends to vary between different manufacturers. Trade requirements might be met if the following mesh size for the various grades could be adopted.

	Mesh Sizes
Coarse Granulated through	6 over 8

	Mesh Sizes
Medium Granulated through	8 over 15
Fine Granulated through	15 over 30
Fine Castor through	30
Medium Castor through	40 over 60
Fine Castor through	60 over 80
Pulverized through	80 over 120
Icing No. 1 through	120 over 200
Icing No. 2 through	200

T.V.P. TEXTURED VEGETABLE PROTEIN

This comparatively new product is made from the soya bean. There are approximately five different sizes available from a fine breadcrumb size to a mince, granule, chunk and strip. These can be obtained in two types – coloured and neutral and in four flavours – pork, ham, bacon and beef.

Its main advantage is to extend meat dishes where it can replace up to 25% in many meat products especially spiced meat dishes like curry.

To prepare T.V.P. it should be soaked in about five times its own weight of water and when hydration is completed the excess water drained off. It requires approximately twice its weight of water for reconstitution.

Since T.V.P. is not meat, it must be ignored in any calculation involving the quantity of meat required by legislation for a particular type of product.

T.V.P. should not be regarded as a substitute but as a complement to meat. In general meat has an excess of fat but T.V.P. has no fat and therefore will absorb this fat. It is claimed that the finished product:

(1) Looks better value and is cheaper.
(2) Has better texture and eye appeal.
(3) Has a better taste, is less fatty and has all the flavour.
(4) The juices of the meat are taken in by the protein product.

YEAST

Two types of yeast are available – Compressed and Dried. The compressed yeast is a grey plastic solid which can easily be broken. Dried yeast is in the form of granules or threads. The fresh compressed yeast only has a limited shelf life and should be used as soon as possible after delivery. Once exposed to the air it soon forms a brown, dry skin in which some of the yeast cells die thereby reducing its fermenting power. Therefore it should be kept wrapped and placed in a cold store (or refrigerator above freezing point) until used.

Dried or desiccated yeast is very useful in hot climates where it is difficult to keep fresh compressed yeast. Several of the yeast cells are killed in the drying process so that it is usual to use the same amount of dried yeast as that of compressed despite the fact that it is more concentrated.

2. *Care of Small Bakery Equipment and Hints to Students*

EQUIPMENT

Obviously to get the best use of equipment, it must be kept in good repair and in a clean condition. The latter is especially important as far as food is concerned. Although this applies to the whole range of equipment used by the baker and confectioner – from an oven to a patty pan – some items merit special attention as follows:

Bain-marie

When in use, the main precaution is to see that the pan is never allowed to burn dry. When not in use, the bain-marie should be emptied, cleaned inside and out, washed in hot detergent water, then rinsed and dried.

Baking Sheets

These can be made in any size but the standard imperial dimensions are 30 in × 18 in. They may have one side open or all four sides turned up. These are made of either aluminium or steel.

Aluminium This has the advantage of being light in weight, rustless and able to conduct heat more quickly. Its main disadvantage is that it is a softer metal and is, therefore, not so durable. It must, therefore, be treated with care. When scraping off food particles never use a steel scraper as this will scratch the surface and may remove some of the metal. The best method of cleaning is to use water with a *mild disinfectant*. If a scraper has to be employed, use either a celluloid or nylon one.

Steel Although these will stand up to harsher treatment, its obvious disadvantage is its ability to rust. If water is used to wash the tin always thoroughly dry afterwards. Often the cause of rust can be traced back to buns being left on the tin to cool. When this happens, the steam trapped between the tin and the bun condenses to water and if left, rusts away that part of the tin under the bun.

Of whatever metal the baking sheet is made, make sure that the corners are really well cleaned. It is here that the particles of food accumulate and if left will bake onto the surface of the tin and then prove difficult to remove. When cleaning baking tins, concentrate on the corners and the centre will clean itself!

It could be thought that the use of silicone paper has removed the need to clean baking sheets. However, if the baking sheet is kept clean the silicone paper can be used for more bakings than would otherwise be the case.

Brushes

These need washing and thorough drying at regular intervals. If left, there is a build-up of food particles adhering to the bristles which quickly reduces the efficiency of the brushes. This is particularly prevalent with board brushes used for brushing such commodities as flour which, when wet, forms a tenacious paste which dries hard onto the bristles.

Mixing Machines and Attachments

These usually have heavy use and breakdown can seriously affect production. Therefore it is well worth while having them regularly serviced and lubricated.

One of the most annoying faults with these machines is the leakage of oil which tends to drip into the mixing. If this happens when sponges or meringues are being made, they could be completely ruined. Cleaning of machines after each use is desirable, but if moving parts like the blade of a mincer are involved, thorough drying is essential to prevent rust formation. Traces of mixtures left on the handles, etc., of the machine are not only unhygienic but can contaminate other mixtures. It is important that these mixers should not be overloaded. This can easily be done by allowing them to mix stiff pastes and doughs.

Always use the correct speed and the appropriate attachment, e.g. a hook for doughs and stiff pastes, etc. Bowls, whisks and beaters, etc., should always be washed in very hot water to eliminate traces of grease which can have disastrous consequences to mixes such as meringues and sponges.

Store bowls inverted so that they do not collect dust and dirt. The best way to store whisks, beaters and hooks is on hooks under a shelf.

Cutters

Pastry cutters are usually very poorly cared for. They are comparatively fragile utensils and with misuse, quickly lose their shape. Always ensure that cutters are put away in their boxes clean and dry. If cutters are replaced in their boxes without proper cleaning, the food particles adhering can go bad without anyone noticing. When cleaning, special attention should be given to the rim where the food particles can easily become lodged. If the cutters are washed, make sure that they are really well dried out before replacing them in their boxes, otherwise rust will soon be apparent. Roller cutters are used for a number of the larger shapes. For their efficiency, besides regular cleaning, an occasional oiling of the bearings is recommended.

Dredgers

These are used for giving a light dusting of flour, icing sugar etc. Since the same kind of dredger is used for these materials, it is important to have some means of identification, otherwise the flour could be mistaken for sugar and vice versa. Regular cleaning is again recommended with a frequent change of contents.

Docker

This is a wooden disc with protruding points used to make a number of small holes in pastry. When cleaning it is essential to remove any food particles from the points.

Roller dockers are now available and their use can greatly speed up this operation. Regular cleaning and the occasional oiling is again recommended for efficiency in use.

Doughnut Fryer

To produce a consistently good product the frying oil should be as fresh as possible and screened free of foreign matter. The fryer should be emptied at regular intervals for cleaning and re-stocking with fresh oil.

Harp

This piece of equipment consists of a thin saw-like blade held in a frame similar to a bow saw. For efficient use the blade should be maintained in a sharp condition by keeping it free from rust and slightly oiled. The blade should be changed as soon as it shows signs of bluntness.

Knives

Besides ensuring that these are kept clean and free from rust, French knives etc. need to be kept sharp for their efficient use. To achieve an edge, a stone should be used first, after which the knife edge should be kept keen by the use of a steel. Stainless steel has an obvious advantage in that it is rustless and is therefore recommended especially for palette knives.

Mixing Bowls

These are of either stainless or tinned steel. Obviously the former has an advantage. Care must be exercised if a tinned bowl needs to be heated since over naked flame there is a danger of the tin-plate melting. The best shape for mixing bowls is spherical.

Moulds

These should be kept in a clean condition both inside and outside, particularly if stacked. Never use a knife to scrape off adhering particles as this will damage the surface and could spoil the mould for future use – especially Easter Egg moulds.

Patty Pans

The cleaning of these is often neglected. Because they are stacked it is very easy to contaminate clean tins with dirty ones. If placed on a dirty tray, dirt becomes transferred to the underside of the tin which if stacked in an unclean condition, makes the clean patty pan into which it nestles also dirty. Always place these patty pans onto a clean tray and ensure they are thoroughly cleaned before stacking.

Provers

These are cabinets in which steam can be generated and in which trays of fermented goods are left to prove. The steam can be made either by allowing a pan of water to boil (gas or electric ring) or by injecting steam from a steam boiler. With the presence of so much water, deterioration due to rust is a real problem in metal provers. Always leave the prover door open after it has been used to allow the interior to dry and so reduce this risk.

Piping Tubes

These should always be washed immediately after use. This is especially important if icings like royal icing are used, because of their rapid drying action. To remove hard sugar from piping tubes, they should be soaked in water, then either brushed out or a thin jet of water directed into the nozzle.

Rolling Pins

The best rolling pins are mounted on ball bearings at the handles and rolling is accomplished by applying pressure and moving the handles away from the person. Obviously in such appliances it is important to prevent the ball race from rusting by the occasional application of lubricating oil. Most rolling pins are made of beech because of its non-splitting properties. The best is undoubtedly lignum vitae, a very hard brown wood. Since all woods tend to warp if left too long in a wet condition, rolling pins should be dried as quickly as possible after washing. On no account should any pastry etc. adhering to the surface of the rolling pin be removed with a knife or similar sharp instrument. This will damage the soft wooden surface of the pin and the impression of this damage would be left upon the paste being rolled out.

Refrigerators

The most common abuse a refrigerator suffers is irregular defrosting. When operating, ice is built up around the evaporating coils inside the refrigerator. This reduces its efficiency and overloads the motor, risking unnecessary breakdowns. Regular defrosting at least once a week should be programmed and the inside wiped over with a weak solution of bicarbonate of soda, then dried before putting the refrigerator back into operation. Notes on the use of Deep Freezers are dealt with on pages 60 and 61.

Siting of refrigerators and deep freeze cabinets should receive careful consideration. They should always be sited in a relatively cool place so that the compressor and its coils can effectively be kept cool. Ideally this piece of equipment is best sited outside the bakery in the open air under suitable cover to prevent ingress of rain.

Savoy Bags and Tubes

Bags are available in either nylon, plastic or cloth. Of these the plastic bags have a definite advantage in that they are not porous and when used for creams etc., no liquid can seep through. Whatever their composition, savoy bags require to be sterilized before each use, especially when used for fresh cream and jelly concoctions – see Food Hygiene. After sterilizing in a suitable solution or by boiling, they should be thoroughly dried and stored in a very clean place. Tubes are made either of plastic or tin-plated metal. In the latter case rust must be guarded against.

Scales

Scales should be regularly serviced by a scale manufacturer to ensure that they always weigh correctly. Careful maintenance and regular cleaning will also help in this respect. Remember that certain weights are legally enforced (e.g. bread) and prosecution can result for short weight. It would be no defence to put blame on a defective scale. Also very small quantities cannot be accurately weighed unless the scales are properly balanced. This could result in more or less of a vital ingredient such as baking powder which could ruin a whole mixing (*see* hints to students on page 41).

Sieves

These may be very fine hair sieves, large meshes for sieving flour and larger still for draining dried fruit after washing.

Sieves should be washed immediately after use. If this is not done, the food or liquid dries in the mesh, making it difficult to remove. To wash, the sieve should be placed upside down under running water, tapping vigorously with the bristles of a stiff brush. Always store sieves in a dry place to prevent risk of rusting wire mesh.

Spatulas and Wooden Spoons

These are usually made of beech and require the same treatment as rolling pins.

Storage Bins

These may be made of metal or plastic, the latter is not as robust as the metal but has the advantage of being more easily cleaned, and are often moulded with no corners in which food particles can be trapped. There should be a regular programme of cleaning. No storage bin should be refilled until it has been thoroughly washed in hot detergent water, rinsed and dried. Storage of such goods as onions, spices etc. should be confined to the same bin since washing is not always effective in removing the pungent aroma of such foods and this can easily be passed on to other commodities via the bin.

Tables

The surfaces of these may be wood, melamine, stainless steel or marble.

Wooden tables can be scrubbed with hot soda water to remove grease, rinsed and wiped dry to prevent undue warping. They must not be scraped with a sharp scraper as lumps might be gouged out. The other surfaces should be washed in hot detergent water, rinsed and dried.

Cutting or chopping should never be done on work surfaces, chopping boards should be supplied for this purpose.

Marble tops require special care to ensure that they do not crack. Heavy concussion such as hitting with a rolling pin can easily cause this to happen.

HINTS FOR STUDENTS

Weighing

Unless automatic balances are used, weighing of small quantities of materials will usually be done on what are termed counter scales. This is where materials are placed in a scale pan and balanced by putting weights on a plate. Some are fitted with a sliding weight which makes it easier to weigh small quantities, and provides less risk of small weights becoming lost or accidentally added to mixtures.

Before weighing on such scales, it should first be made to balance. This will ensure that the wrong scale pan is not mistakenly used (if there is more than one pair of scales). It will also indicate if dried food, material or dirt has adhered to either the scale pan or the weight platform, and so altered the balance.

Having ensured before the weighing operation that the scales balance perfectly, also ensure that the correct amount of material is weighed by observing that the scales balance again (not down on one side). Such accuracy is very important in the weighing of very small quantities of vital ingredients such as baking powder.

Never attempt to guess at weights even if these are very small. It is impracticable to attempt to weigh a quantity of less than 5 g or $\frac{1}{4}$ oz.

If it is a powder, quantities less than this can be weighed as follows:

Let us suppose we require $\frac{1}{16}$ oz. The best method is first to weigh $\frac{1}{4}$ oz, place this quantity onto paper and with a knife or scraper physically divide into 4 and use a quarter part.

Always endeavour to keep the scale pan as clean as possible. Fats for example should be weighed onto either the sugar or the flour of the recipe and not on its own. This would unnecessarily soil the scale pan which would require washing in hot detergent water before other materials could be weighed.

For liquids it is best to weigh a container first full and then weigh the liquid by pouring it into the container on the scale pan. Critical amounts of liquids should always be weighed, never measured in a calibrated vessel like a quart pot.

Preparation for Mixing

Before a machine mixing is switched on, ensure that all the materials are correctly weighed into bowls, basins or on papered trays. Also check that the necessary tins, moulds etc. are prepared. This will enable the operator to concentrate on the job of mixing. Attention to details other than to the actual mixing may result in overmixing or the wrong machine speed being used resulting in goods spoiling.

Always use a plastic scraper (the rounded side) to keep the sides of the mixing bowl free of batter so that a homogeneous mixing results. The plastic scraper can also be used like a spatula for the hand-mixing of such goods as macaroons where they automatically clear the batter as it is mixed.

Savoy Bags

When a savoy bag is filled with a mixture the top should be turned over to prevent the batter from soiling the top. If the last portion of a mixture needs to be scraped into the savoy bag, and there is no one to assist, the bag can be held over the edge of the table with a heavy weight, leaving both hands free, one to hold the bowl and the other the scraper.

If the mixture is very soft or liquid, the portion of bag nearest the tube should be given a twist and pushed into the tube to form a plug before the bag is filled.

If the mixture is very stiff, considerable energy will be expended if pressure is brought to bear on the top of the mixture in order to force it through the tube. This is overcome if a twist is made in the bag half way down, then, whilst one hand keeps the top of the bag closed to prevent the mixture from being squeezed out, the other hand can squeeze the remainder out of the tube at the lower end of the bag without so much effort.

If expensive ingredients are used, such as almond mixtures, it is desirable to scrape the inside of the savoy bag and extract the maximum amount of mixture. This can be done with the end of a palette knife by scraping the mixture down into the tube.

Where two differently sized tubes are required for the same mixture, e.g. meringue shapes, it is not always necessary to change the savoy bag. Provided the larger tube is fitted, all that is required is to hold the smaller tube over the larger in order to pipe out the required smaller shapes.

Once a savoy bag has been used the tube should be removed, the bag thoroughly washed in hot detergent water, and hung up to dry.

Cutters

When cutting shapes out of pastry, for lining patty tins, dust the table liberally to prevent the pastry sticking, but brush off the flour from its top surface before cutting out. Then invert the pieces before placing them into the tins. The sticky side then adheres to the tin whilst the floured surface makes thumbing easier.

Baking

Many goods are ruined because of the wrong baking conditions. If uncertain, goods baking in an oven should be inspected during the baking period and remedial action taken if necessary. Too much bottom heat can be minimized by double traying. Excessive colouring of a cake top due to too much top heat can be reduced by covering with a sheet of paper. Whatever precautions are taken to reduce the effect of excessive heat on baked goods, it is essential that they remain in the oven until they are cooked. The results of undercooking produce a fault more serious than a high crust colour.

Washing Utensils

There is always a great temptation to place dirty utensils in a sink with the idea of a grand washing-up at the end of the day's production. This should be resisted since it is far easier to wash tools and utensils as soon as they have been used rather than leaving them for a few hours before cleaning. Once the residues of mixtures are left on utensils to dry, they are difficult to clean and it requires twice as much effort and time to remove the particles. The golden advice is – Wash each utensil as it is used!

BREAKING OPEN MATERIALS IN BULK

Before materials are used it is necessary to break open the box or packaging in which they are sold and stored. Often the raw materials are not removed from their packaging before quantities are removed for weighing into recipes. Such a practice should be discouraged for hygienic and other reasons which will become apparent later.

Fats/Margarine etc.

The cardboard box of fat should be opened and the contents removed. Today fats and margarines are usually packed in a plastic lining so that removal together with lining is an easy operation.

Have available a plastic tub container or tray into which the fat can be transferred.

Carefully remove the plastic liner. This is made easier if the fat is refrigerated or very cold. If it is in this condition the lining can be peeled away easily, but if the fat is at a temperature at which it is still soft, a plastic scraper may have to be used to remove the adhering fat.

Marzipan/Almond or Sugar Paste

Because only small amounts of this material are used at a time and because it forms a skin when exposed to the air, it is customary to keep it in the pack in which it is delivered, since the material is usually wrapped in a plastic liner. However, prolonged storage even in such a liner will not prevent the paste from forming a skin and if handfuls are dug out of the mass, such skin is difficult to remove without also removing much of the paste. Whenever any paste is removed from a bulk quantity, it should first be emptied out of its container, and the plastic liner carefully furled back. The required quantity of paste should then be cut off cleanly with a knife so that a flat surface is left before wrapping and re-placing in the box. If the cut surface skins it is a comparatively easy matter for this to be removed with a sharp knife.

Rolling out Pastry or Dough

Students rolling out pastry or dough for the first time always experience difficulty in shaping a rectangle, square or circle.

Circle Mould the paste or dough into a ball and extend the size by rotating the ball and at the same time rolling out with the rolling pin in every direction to the size required.

Rectangle or Square (*see* Figure 1 below) To obtain a slab of paste or dough with straight sides and square corners first roll out the ball in opposite directions in the *centre* only (*see* Stage 1). This will leave it thicker at the corners. By rolling these edges out with the rolling pin at an angle of 45°, we can achieve a square corner before extending to the final size required (*see* Stage 2).

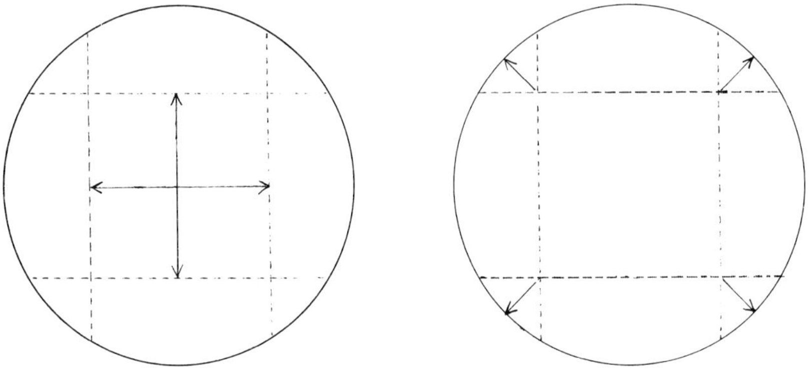

Figure 1. Stages in rolling out pastry

Stage 1 Stage 2

3. Breadmaking Technology

It is not the purpose of this essentially practical craft book to write a detailed technical thesis on fermentation and its utilization to produce first-class bread. But without some essential relevant information, the craftsman will lack the understanding necessary to produce first-class goods, and therefore a brief summary of the materials and factors governing fermentation is introduced here.

Yeast

The agent responsible for the fermentation of bread and fermented goods is yeast. This is a living micro-organism existing as a single cell having a diameter of $\frac{1}{100}$ to $\frac{1}{160}$ mm ($\frac{1}{2500}$ to $\frac{1}{4000}$ in) invisible to the naked eye, but easily discernible under the microscope.

When introduced to warmth, moisture and food, this organism ferments, i.e. it produces carbon dioxide gas and ethyl alcohol and at the same time reproduces itself. (Under the last condition one cell dividing into two every 20 minutes.) It is the carbon dioxide gas generated by the yeast with which we are concerned in breadmaking, although the ethyl alcohol contributes to the flavour in the finished goods.

The whole process of yeast fermentation is very complicated and there are many factors which affected it. Therefore the best way to understand these factors is to elaborate upon each.

Food Material

Most flours contain all the food necessary for the fermentation of yeast. The most essential food is dextrose sugar (glucose) which is directly fermented into carbon dioxide by a group of enzymes in yeast called zymase. Because yeast contains enzymes which are capable of changing cane sugar (sucrose) into fructose and glucose, and malt sugar (maltose) into glucose, almost any sweet material (except milk sugar) will act as a source of food material.

Wheat contains about 2·5% of sugar made up of 2% sucrose and ·5% maltose and so it will readily ferment without the aid of extra sugar, provided a short process is employed. For long processes, i.e. overnight, or for flour naturally deficient in sugar (low maltose content) there could be a slowing down of the fermentation due to the yeast utilizing all the available sugar. In these circumstances the use of a source of sugar in the dough is not only essential to stimulate the yeast activity but also helps to impart colour and bloom into the crust and crumb.

The concentration of sugar present which yeast is expected to ferment is very important. The optimum is about 12½% and concentrations above this have a retarding influence. Thus, for example, in bun doughs with a sugar content considerably higher than this, extra yeast has to be used to compensate for this effect.

In addition to sugar, yeast also requires certain soluble inorganic salts and ones which contain nitrogen which is essential to growth. These are provided in proprietary products which although in reality are yeast foods, are usually called *improvers*. One mineral used for this purpose is ammonium chloride which is recommended to be used at the rate of 45 g per 100 kilos flour or 2 oz per sack (280 lb) flour. This agent is invariably found in most proprietary improvers.

Salt, fat and spices all have a retarding effect upon yeast fermentation and if these are used in excess a higher proportion of yeast must be used.

KNOW YOUR WHEAT GRAIN

A grain of wheat cut lengthwise (through crease)

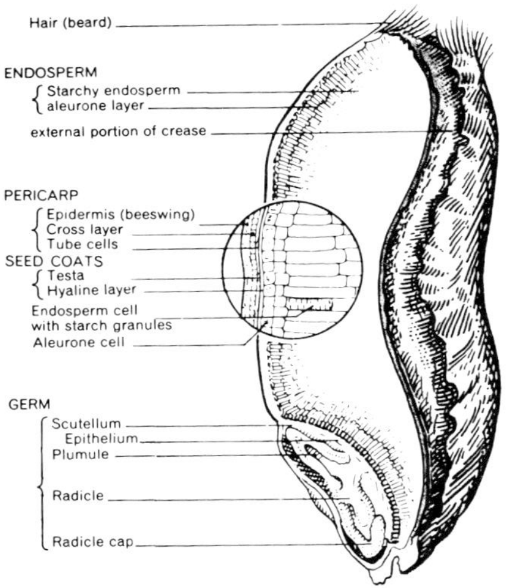

Whole grain of wheat

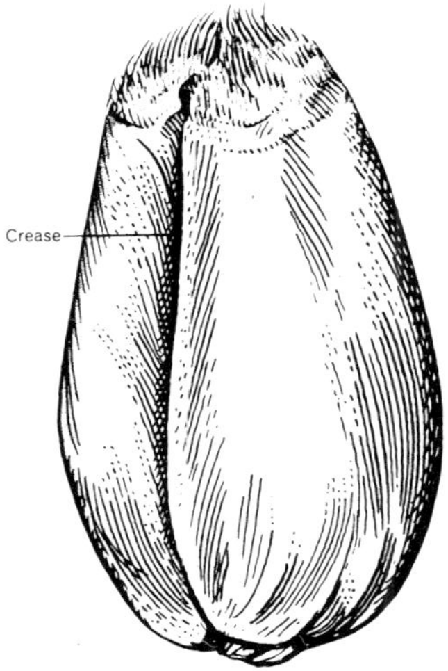

A grain of wheat cut across the middle

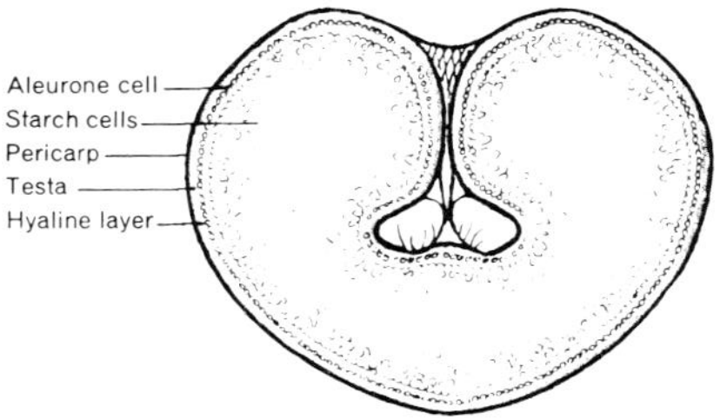

A Seed

The wheat grain as a seed is fitted for reproducing the plant from which it came. The *germ* is an embryo plant, with a radicle which can grow into a root system and a plumule which can develop into stems, leaves and ears. The *pericarp* is a tough skin which protects the inner seed from soil organisms which may attack it. The inner *seed coats* control the intake of water by the seed. The *endosperm* is the food reserve on which the young plant lives until it has developed a root system.

Which is Milled

The purpose of milling is to reduce the wheat grain to a fine powder *flour*. A single grain makes about 20,000 particles of flour. In *wholemeal flour* all parts of the grain are included, but in producing *white flour* the seed coats and the embryo are not used. Instead, they are flattened and removed as small flakes, by sifting over nylon or silk mesh. These flakes are referred to collectively as *wheatfeed.*

For Our Nourishment

The flour which comes from the grain of wheat is used in making bread and biscuits, cakes and confectionery, puddings and pies. This wheaten flour is rich in carbohydrates (for energy), protein (for growth and development), the essential **B** vitamins (for good health, good nerves and good digestion) and important minerals like iron (for healthy blood) and calcium (for strong bones and teeth).

Flour gives us dishes that are good to eat – and the nourishment essential to good health. What is more, the seedcoats, or *wheatfeed*, not used in making white flour are valuable food for livestock, and so help to provide us with eggs, bacon, meat and milk.

Flour

We require a medium to strong flour for breadmaking, i.e. a high % of gluten.* Average home-milled breadmaking flours have a gluten content of about 11–12%. Stronger flours than this are available milled either in this country from strong wheat or are imported from Canada.

Another quality required of a breadmaking flour is the maltose figure. Besides the sugar naturally occurring in flour, there are certain diastatic enzymes present which transform the degraded starch cells in the flour into maltose sugar and dextrin gum. The use of a flour with a high maltose figure will result in bread which has a high crust colour and sticky crumb. If flour has a very low maltose figure some sugar additive should be used, otherwise the crust colour will be anaemic and in long processes there is the possibility that the fermentation will slow as the yeast utilizes all the available sugar.

The colour of the crumb of bread is controlled to some extent by the colour of the flour used, but it can also be influenced by the manipulation of the dough (see under Moulding, page 54).

ADDITIVES

Fat

Any type of fat or oil can be used although butter and lard will obviously give a better flavour. The advantages claimed by the use of fat are as follows:

(1) Gives an increased volume to the bread.
(2) Produces a more tender and thinner crust.
(3) Improves the colour of the crumb.
(4) Softens the crumb.
(5) Reduces toughening in milk bread.
(6) Increases energy value.
(7) Improves flavour if butter or lard is used.

Fat can be used at the rate of approximately 1·5% (15 g per kilo) of the weight of flour, but usually half this quantity is sufficient to bring about a marked improvement.

Crumb Softeners

As the name suggests, these agents bring about a more marked improvement than fat in the softness of the crumb in bread. There are four in wide use today, some as a component in proprietary improvers. They are as follows:

(*a*) *Glyceryl Monostearate*–This is a cream-coloured, free-flowing powder or flaked product which should be added at the rate of 90 g per 100 kg flour (3–4 oz per sack). Before use, however, it must be made into an emulsion, one part being whisked with five parts of hot water.

(*b*) *Stearyl Tartrate*–This is another emulsifying agent similar to glyceryl monostearate but used at only half the quantity.

(*c*) *Lecithin*–Another powerful emulsifying agent which when used at the rate of 90–180 g per 100 kilos flour or 4–8 oz per sack, reduces stickiness in a dough and allows more water to be used. It is often used by mixing it with fat and flour as a premix in the proportions of lecithin 11 parts, shortening 4 parts, flour 5 parts. This can then be incorporated at the rate of up to 1% of the flour quantity.

(*d*) *Soya*–*see* below.

Soya

This is a very useful bread additive and the following advantages are claimed for its use.

*The name given to the protein content of flour.

(1) Softens the crumb.
(2) Enables more water to be used.
(3) Natural enzymes present have a bleaching effect.
(4) Improved appearance and colour of the crust.
(5) Slightly increases the protein content of the bread.

It is used at the rate of 70 g per 100 kg flour (2 lb per sack) usually mixed first with an equal weight of water, but it can also be added dry or mixed with fat.

Sugar

Any sugar may be added to bread but in practice we usually use the following:
Granulated, castor or syrup This should first be dissolved into some of the dough water (liquid). Ensure that it is not brought into direct contact with the yeast.
Fondant Because this contains invert sugar it is more quickly fermented and the bloom of the crust and crumb is improved. For exhibition bread, a mixture of fondant and lard in equal proportions rubbed into the dough at the knock-back stage is claimed to bring about a marked improvement in the bread produced.

We can now summarize the use of sugar in fermented goods as follows:

In bread –
- (*a*) To increase gas production
- (*b*) To impart colour and bloom to the crust.

In fermented goods –
- (*a*) As a sweetening agent.
- (*b*) To dress the tops of buns, etc. for decoration.

Malt

Available as either high, low or non-diastatic and their uses must be well understood if distastrous results are to be avoided.
High-Diastatic Malt These malts are very high in enzyme activity being capable of breaking down the ruptured starch cells in flour to produce maltose sugar and dextrin gum. This occurs at temperatures beyond that at which yeast is killed, therefore more sugar is produced than can be utilized by the yeast in fermentation. Used with the wrong flour and process therefore, they will produce bread showing the characteristics of a high-maltose flour with the sticky crumb and high crust colour. *The use of this type of malt is beneficial with short-process doughs when a very strong flour is used.* This is because of the action of the proteolytic enzymes in the malt which can soften the gluten and therefore produce a softer crumb. These malts are pale in colour.
Low-Diastatic Malt The use of this will provide the yeast with more fermentable sugar with less of the enzyme effect of the former. The softening effects will, therefore, be less. This malt is cured at a higher temperature and is therefore a darker colour.
Non-Diastatic Malt There is no enzyme activity with these malts because they have been kilned at a high temperature, resulting in a darker colour and a more pronounced flavour. Their use will only provide a source of fermentable sugar and flavour.

The following advantages are claimed from the correct use of malt:
(1) Aids the thorough ripening of the dough to give an increase in volume and gas production.
(2) Improves the crust colour due to the extra sugar produced.
(3) The bloom is improved because of better fermentation.
(4) A softer and lighter crumb is also produced because of more thorough fermentation.
(5) Flavour is improved.
(6) Improves the keeping properties.

Types and Quantities There are three types of malt products – malt flour, viscous malt extract and dried malt extract, and they should be used in the following quantities:

Dried malt extract:	Up to 250 g per 100 kg flour (12 oz per sack flour).
Malt extract:	Up to 715 g per 100 kg flour (2 lb per sack flour)
Malt flour:	Up to 1 kg 430 g per 100 kg flour (4 lb per sack flour)

Malt Bread

For malt bread we use considerably higher quantities of malt than that used as an improver for white bread. The special consideration that this involves will be dealt with separately under Malt Bread (*see* page 92).

Fungal Enzymes

These are destroyed at temperatures lower than the cereal enzymes in malt and therefore are to be preferred as a corrective additive for use with a low-maltose flour. The use of this agent is mostly confined to millers who treat the flour before selling it to the customer.

Salt (Sodium Chloride)

Principally this is used to impart flavour into what would otherwise be an insipid-tasting foodstuff. Apart from enhancing flavour it has other side effects as follows:

(*a*) Acts as a stabilizer by strengthening the gluten of the dough.
(*b*) Improves the colour and bloom to both the crust and the crumb.
(*c*) Reduces staling.
(*d*) Retards fermentation.

This last effect must be considered when making doughs to see that the yeast and the salt do not come into close contact with each other. The concentration of salt in a dough is tolerated by the yeast, provided that at the mixing stage it is kept in a sufficiently dilute solution. Salt may be added either with some water at the mixing stage or dusted dry into the dough as it is being made. In high-speed mixers it is just added dry along with all the other ingredients. For long fermentation doughs, an increased quantity of salt is often used especially in Scotland as this helps to control the fermentation.

Quantities in Use

This can vary according to local preferences. Typical amounts are as follows:—

Low	: 1 kg 420 g per 100 kg flour (4 lb per sack flour)
High	: 2 kg 500 g per 100 kg flour (7 lb per sack flour)
Average	: 1 kg 785 g per 100 kg flour (5 lb per sack flour)

Other ingredients are dealt with in the following sections:—
Gluten – *see* page 94.
Milk – *see* page 75.
Mould and Rope inhibitors – *see* pages 59 and 60.
Fruit, nuts, spices, etc. – *see* Chapter 8 page 108.

THE BREADMAKING PROCESS

Basically bread is made from a mixture of four ingredients – flour, yeast, salt and water in the correct quantities. This is usually left for a period of time and at a certain temperature to ferment and be brought to the correct condition from which good bread can be produced. Before dealing with the individual process we ought to explain the terms *temperature* and *time* in relation to fermentation.

TEMPERATURE

At 0°C (32°F) yeast is dormant but as the temperature increases so too does the activity of the yeast up to a maximum of about 35°C (95°F). At about 44°C (110°F) the yeast cells become killed and fermentation ceases. Since extreme cold does not entirely kill yeast, it makes it possible for unbaked doughs to be deep-frozen and there is now a very large market in this commodity.

The best working temperatures for breadmaking are between 21–29°C (70–85°F) although it is common practice in plant bakeries at the final stages of proving to increase the temperature to 38°C (100°F) for the maximum gas production in tin bread.

For doughs made on a no-time dough process, higher temperatures of approx. (32°C) 90°F are often used to stimulate gas production and the rapid ripening of the dough. Too high a bulk fermentation temperature encourages undesirable bacterial growth and increases dough-skinning unless adequately covered.

To determine the temperature of the water required to make a dough at a given temperature, there are two methods:

1. *Simple Method.*
   (i) Determine the temperature of the flour to be used.
   (ii) Double the required dough temperature and subtract the flour temperature.
   (iii) The result is the temperature of the water required.

Example Let us suppose that we want a dough at 27°C (80°F) and the flour temperature is 16°C (60°F) then:

(Dough temp. × 2) – flour temp. = water temp.
27°C (80°F) × 2 – 16°C (60°F)
54 (160°F) – 16°C (60°F) = 38°C (100°F)

The temperature of the water required to make a dough at 27°C (80°F) using flour at 16°C (60°F) is therefore *38°C (100°F)*.

This method works reasonably well with large doughs which retain their heat but with small doughs, especially if made in cool conditions, it might be necessary to take the water a few degrees hotter.

2. *Major Factor Method.*

The former method only takes into consideration the temperature of the flour whilst in the second method the temperature of the baking or mixing room is also considered.

Let us suppose that we have made a dough successfully at the temperature required. We now calculate a factor by adding together the temperatures of the flour, water and bakery. This figure called the *major factor* may now be used to make subsequent doughs and because the bakery temperature fluctuations are taken into consideration, the final dough temperatures of subsequent batches are more likely to be consistent.

Example Let us assume that when we made the dough by the simple method, the bakery temperature was 21°C (70°F). Our major factor now becomes:

	°C	°F
Bakery temperature	21	70
Flour temperature	16	60
Water temperature	38	100
	75	230

If on the following day, the bakery temperature dropped to 16°C (60°F) and the flour to 12°C (55°F), we have only to take these figures from the major factor to ascertain the temperature at which the water for the dough needs to be taken.

Thus:—

	°C	°F
Major factor	75	230
less bakery temperature	16	60
less flour temperature	12	55
Water temperature required	47	115

CALCULATION OF WATER TEMPERATURES FOR SPONGES AND BATTERS

Water temperature calculations of doughs are comparatively easy because of the fortunate chance that the specific heat of flour which is approx. 0·5 is balanced by the fact that there is approx. twice the quantity of flour than water $0{\cdot}5 \times 2 = 1$. In mixtures where we have an inbalance, however, we have to rely upon calculations involving heat transfer between the water and the flour. An example of such a calculation is the determination of the temperature of water required to make a sponge (*see* page 66) at 27°C (80°F) as follows:

Example (using Metric Scale) A sponge is required to be made at 27°C from 100 kg of flour at 21°C and 110 kg of water. At what temperature will we require the water?

(*a*) Heat gained from the flour will be

Weight × Specific Heat × Temperature Rise

100 $\frac{1}{2}$ 6

(*b*) This heat has to be gained from the water which will lose

Weight × Specific Heat × Temperature Loss

110 1 (water temp. − sponge temp.)

(water temp. − 27)

The gain at (*a*) must equal the loss at (*b*)

Therefore:—

110 × (water temp. − 27)	$= 100 \times \frac{1}{2} \times 6$
110 × water temp. − 2970	$= 50 \times 6$
110 × water temp.	$= 2970 + 300$
Water temp.	$= 3270 \div 110$
Required water temp. is therefore	= 29.72
or in practice	30°C

Time

When yeast commences to ferment, a very complicated series of enzymic changes takes place not only producing carbon dioxide gas and alcohol, but also bringing about a modification of the gluten making it soft and more extensible. This enables the gluten network in the dough to stretch further and so hold more gas. A perfect dough is one in which this function has been allowed to reach the optimum, i.e. the time allowed for the dough to ferment in bulk has been correct for the temperature chosen and amount of yeast used. Below this optimum the dough is said to be *under-ripe* and above this optimum, the dough is said to be *over-ripe*.

Therefore, for a perfect dough to be made, there must be the correct correlation between the dough temperature, dough time and the yeast quantity employed. If the dough temperature is reduced, the yeast quantity must be increased and vice versa. Likewise the longer the process, the less yeast which can be used because of its own growth in the fermenting dough.

Note In mechanically developed dough, such as that used in the Chorleywood Bread Process, this ripening is achieved by an input of work and since it is a no-time dough we rely only upon the yeast to provide the gas for aeration.

Yeast Quantities

The amount of yeast required for any dough is difficult to assess accurately and is usually arrived at by a knowledge of the conditions by which the bread is to be made, the ingredients used, the temperature and the time allowed for bulk fermentation. There is, however, a formula for the latter based upon a numerical factor which is then divided by the time in hours. This has been arrived at by trial and error but works very well for short-process doughs.

The basis for this factor is that if we take a typical bread dough based upon a sack quantity (i.e. 280 lb) in the Imperial system of weights, we get the following:

Dough at 27°C (80°F) for 3 hours
Yeast quantity required = 3 lb

The factor is found by multiplying the yeast quantity by the time, i.e. $3 \times 3 = 9$.

Example Let us assume that we require a dough at 27°C (80°F) for two hours bulk fermentation –

Yeast quantity required $= 9 \div 2 = 4\frac{1}{2}$ lb

If we now turn to the metric scale and base our yeast quantity on 100 kilos, we get:

3 lb of yeast per 280 lb flour is approx. equal to 1%.

Therefore, a 3 hour dough at 27°C (80°F) will require 1 kg of yeast per 100 kg flour.

The factor therefore becomes $1 \times 3 = 3$.

Thus, if we require a dough on a 2 hour process, the quantity of yeast we would require per 100 kg flour

$= 3 \div 2 = 1$ k 500 g.

There are exceptions even to the above rule. These are as follows:

(*a*) If the time between the stage at which the bread is taken for moulding until the time when it is actually baked is prolonged, i.e. because of handmoulding – this factor must be reduced by approx. 10%.

(*b*) For fully automatic plants this factor should be increased by approx. 10%.

(*c*) For a no-time dough the yeast quantity should be approx. $2\frac{1}{2}$% of flour weight.

(*d*) One hour doughs require a yeast quantity of approx. 2% of flour weight.

(*e*) For small doughs a higher factor must be used to compensate for heat loss during fermentation. An approximate figure for recipes in this book based upon the 1 kilo ($2\frac{1}{4}$ lb) of flour is as follows:

Metric	4·2 kg per 100 kg
Imperial	12 lb per sack (280 lb).

To operate this formula for these small quantities –

For a 3 hour dough at 26°C we require $\frac{4 \cdot 2}{3} \div \frac{100}{1} = \cdot 014$ kg

Per 1 kg flour we require $= 14$ g yeast

For a 3 hour dough at 80°F we require $\frac{12}{3} \div \frac{280}{1} = \cdot 032$ lb

Per $2\frac{1}{4}$ lb flour we require $= \frac{1}{2}$ oz yeast.

Dough Temperature Variations

Temperatures cooler than the example given, progressively reduce the rate of fermentation so that at 5°C (40°F) it is practically dormant. Obviously therefore, we need to know how much adjustment we have to make to the yeast quantity to make a dough at any given temperature.

It has been reliably established that a difference of 1.1°C or (2°F) requires an average of 10% more in bulk fermentation time.

Example A two-hour dough calculated to be at 27°C (80°F) was found to be at 24°C (76°F) Such a dough would need to be left another 12 minutes.

For cold doughs the accepted rule is to progressively increase the yeast until at 21°C (70°F) the quantity is double that which would have been required for a dough of the same size but at 27°C (80°F). The inter-relationship between yeast quantity and dough temperatures may be shown thus:

Dough Temperature °C	°F	Percentage increase on quantity of yeast used.
26·8	80	—
25·5	78	+20%
24·4	76	+40%
23·3	74	+60%
22·2	72	+80%
21·0	70	+100%

Water Quantity

The consistency of a dough will obviously depend upon the type of bread being produced, the strength and type of flour used, the length of process, the influence of additives and whether or not it is made on a high-speed mixer.

Type of Bread

Obviously if bread is to be baked in a pan or tin which supports its shape, a much softer dough containing more water can be made. Bread which is baked without a pan, i.e. coburgs, cottage or bloomer loaves, needs to be made with less water in order to produce a dough which will support the shape into which it is moulded.

Doughs which have to be used for making harvest festival loaves, e.g. wheatsheaf, would need to be made very stiff with a greatly reduced water content.

Malt bread needs to be of a tight consistency when made, to help minimize the excessive softening which takes place during the fermentation of this type of bread.

Wholemeal bread is also usually made slightly on the stiff side. Although more water is often used than for a comparable white dough, a considerable amount is held by the bran present.

In some fermented products the finished dough resembles a batter, as for example in crumpets, which are poured onto a hotplate to bake.

The water content influences bread in several ways as follows:

Shelf Life

The more water present the longer it takes for the loaf to lose its freshness hence the longer it keeps moist.

Mould

Wrapped bread with a high moisture content is more prone to go mouldy and mould inhibitors (*see* page 24) are widely used by the large producers to help minimize this effect.

Crust

In unwrapped bread, the crust of bread made from tight doughs cracks into large pieces whilst from slack doughs, the crust crazes into many more small pieces.

In wrapped bread, the crust of high moisture content loaves rapidly goes leathery as it absorbs the internal moisture of the loaf.

Flour

The water absorption of flour is to a large extent influenced by its strength, the gluten for strong flours requiring considerably more water for its hydration during doughmaking than that of weak flours.

Length of Process

During fermentation there is a softening of the dough and this can only be compensated for by reducing the water content and making the dough tight initially. Long fermentation processes are now out of favour because a considerable loss of yield results due to this reduction of moisture. No-time and very short processes can use doughs which can carry greatly increased moisture content.

Use of Additives

There are several additives, e.g. soya, which are capable of absorbing moisture and their use can increase the amount of water used in the dough and hence the yield.

Milk powder is another agent which brings about an astringent effect and therefore demands a higher water content in the dough.

Chorleywood and Other High-Speed Processes

This utilizes a no-time dough process, additives and high speed mixing to produce a dough with a very high water content up to 64% based upon the flour weight.

SUMMARY OF THE BREADMAKING PROCESS

Doughmaking

The materials in the recipe are first weighed and then mixed to a dough. Whether this is done by hand or machine the aim should be to produce a clear elastic dough free from any stickiness. During the mixing of the flour and water, the gluten is first formed and then developed. During this stage, the gluten imbibes water to become fully hydrated, a process which is essential if bread of good volume is to result.

Care should be taken to ensure that the yeast and salt are separately mixed into the dough.

Bulk Fermentation

The dough is now set aside, covered to prevent skinning, for a period of time, called the bulk fermentation time (B.F.T.) which is taken from the time from which the dough is actually made until it is taken for weighing, (scaling), moulding and proving. Ideally, the dough should be allowed to ferment at this stage at a temperature approx. to that at which it was made. If left in colder conditions or in the presence of draughts it will become chilled, the fermentation rate will be reduced and under-ripeness will result. Conversely, if left in too warm an environment, fermentation will proceed at too fast a rate causing over-ripeness.

Knocking Back

This is carried out at about $\frac{3}{4}$ or $\frac{2}{3}$ of the B.F.T. It consists of re-mixing the dough for a few minutes. This achieves three objectives:

(1) It expels the gas thus reducing the volume of dough.
(2) Equalizes any unequal temperatures in the dough.
(3) Has a beneficial action upon the gluten structure in the dough by the stretching action involved.

Whilst it is possible to make good bread without knocking back, this process certainly improves the quality of the bread produced.

Scaling

At this stage the dough is divided into pieces ready to be individually moulded into shapes. When scaled by hand it is weighed into the appropriately sized pieces according to the legal requirements. Since there is approx. a 10–12$\frac{1}{2}$% loss of water due to its transformation to steam in the oven, the dough has to be scaled at a weight greater than that required of the bread, in order to compensate for this loss. A loaf to retail at 800 g (28 oz) has to be weighed at 910 g (32 oz).

If divided by machine, the pieces are divided by volume and not by weight, although check weighing has to be carried out to ensure that the legal requirement is met.

Bread which requires a longer baking time, such as crusty bread, may have to be scaled at an even greater weight to compensate for the loss of moisture during baking. For plant tin bread, baked in the minimum time, the weights above can be slightly reduced.

Handing Up

This is the preliminary moulding stage at which the dough pieces are moulded to a round ball shape. They should have a fine unbroken and smooth skin.

Intermediate Proof

The pieces are now given approx. 12–15 minutes rest in which to recover and regain their extensibility before being moulded into their final shape.

Final Moulding

The shape of the finished bread as well as its texture is largely governed by the final moulding it receives. Hand-moulding can often produce better results than a machine because the hand is sensitive to the pressure required to bring the dough piece to the desired shape. The dough must first be flattened to expel the gas and then moulded into the required shape without trapping any gas which will result in holes in the texture and a poor shape. Great care has to be taken with moulding machines to ensure that pressure plates, etc., are so adjusted as to neither tear the dough nor mould it too loosely. Both faults will greatly affect the crumb structure of the dough. A perfectly moulded dough piece should not only have a good shape but have a smooth unbroken skin.

In many bakeries tin bread is made from several pieces of dough in what is a multipiece loaf. This can either be four pieces placed concertina-like into the tin or two long pieces twisted. This produces a much strengthened crumb which has good buttering ability. The formation of a finer and smaller cell structure at the surface also improves the colour as less light is absorbed.

Final Proof

Final moulding should have resulted in the dough piece being degassed and the gluten knitted together to form a compact piece of dough. The aim of the final proof is now to allow a uniform expansion of this dough piece to take place prior to it being baked. To prevent skinning this should be in a humid atmosphere either in a prover in which these conditions are generated, or by being kept well covered. The temperature can also be increased.

The final proof-time varies considerably according to the process used, the type of goods being made and the temperature. It can vary from 30 minutes for no-time doughs made and proved at a high temperature to 1 hour 45 minutes for longer process cool doughs from which exhibition bread can be produced.

Correct proof is essential to produce a loaf which is of good volume but has a stable texture (*see* Faults, pages 62 and 63).

Accurately judging final proof can only be acquired with experience. Not only is final proof judged by the volume of the dough piece, but also when the surface is gently pressed with the fingers, it should spring back. If the impression of the fingers remains, it is a sign that the dough is either over-ripe or over-proved. Also, excessive steam should be avoided as this too weakens the dough surface.

Baking

The temperature range for bread is between 232°C–260°C (450°–500°F) with a good average being 246°C (475°F). The baking of bread extracts a lot of heat from the oven and it is often the practice of bakers to load bread into an oven hotter than required because of the fall in temperature which occurs afterwards. The time of baking depends upon the size of the loaf and the volume being baked at one time. As a guide–

a 400 g (14 oz) loaf requires approx. 30–35 mins.
an 800 g (28 oz) loaf requires approx. 35–40 mins.
a 1600 g (56 oz) loaf requires approx. 40–50 mins.

A good method to test whether bread is baked is to tap the underside. If it is thoroughly cooked it will give a hollow echo.

Despite these high temperatures it is an interesting fact that the centre of the loaf rarely reaches the boiling point of water. The time taken for bread to be adequately baked varies according to its size, the type of oven and whether the oven is fully loaded or not. In some ovens it is possible to bake bread in under 30 minutes for a 400 g (14 oz) loaf whilst Scotch batch bread, for example, may take as long as $1\frac{1}{2}$ hours because the loaves are so solidly packed into the oven.

Pans

The type of bread tin or pan used will also affect the baking time. Solidly constructed tins will take longer for the heat to penetrate than ones of a lighter construction. Aluminium tins will conduct heat quicker than steel and have the advantage of being lighter in weight. Bread tins should have a dulled surface to prevent heat being radiated away from the loaf during baking. New tins which have a shiny surface should be placed in an oven at approx. 204°C (400°F) for at least 3 hours for its surface to become dulled before being put to use.

CHANGES WHICH TAKE PLACE IN BAKING

Once the dough is subject to the heat of the oven a series of changes takes place. Firstly there is expansion of the gas already in the dough piece followed by an increase of gas production by the yeast as the temperature rises. Water vapour and ethyl alcohol may also slightly contribute to this expansion. At approximately 43°C (110°F) the yeast becomes inactive and is eventually killed at 54°C (130°F). By this time the final volume of the loaf is more or less fixed, the difference between the volume before baking and after baking being called *oven spring*. Gelatinization of the starch occurs about 65°C (199°F) some being converted to maltose and dextrin by the diastatic enzymes present (*see* Malt, page 47). This is followed by the coagulation of the gluten which begins at 74°C (165°F) by which time the loaf has set. Crust formation occurs as the outside of the loaf reaches the high oven temperature and gradually a brown colour is formed from the combination of sugars and dextrins which form on the surface. The formation of the latter is assisted by steam and results in the production of a glaze. A loaf of bread is usually judged to be baked by having a satisfactory crust colour and also a hollow sound when tapped with the knuckles.

Steam

The use of steam in the oven is essential for well volumed symmetrical loaves to be produced. There are four main advantages:—

(1) It prevents the premature drying out of the crust so that expansion of the dough piece can be unrestricted and uniform, giving a good shape. For oven-bottom bread the use of steam for this purpose is essential.
(2) The excessive evaporation of water from the loaf is prevented thereby helping to reduce loss of dough weight.
(3) Assists in the distribution of heat in the oven.
(4) Produces a glaze on the crust.

Excessive use of steam will produce a tough and leathery crust and must be avoided.

Steam is usually injected into an oven from a special boiler designed to produce either "dry" or "wet" steam, the latter containing droplets of water. In the first few minutes of a loaf entering the oven, the steam condenses onto its surface immediately gelatinizing the starch and producing the dextrins which form the glaze.

Cooling

Before bread can be sliced and wrapped it must be cooled so that the internal temperature is lowered to approx. 27°C (80°F). This can be achieved by stacking the bread in such a way that cool air can circulate around it. In automatic plants special conveyor belts carry the bread through an air-conditioned chamber but the principle is the same. When properly cooled it should be possible to produce a stable clean, crumb-free slice which can be easily buttered.

Staling

A considerable amount of research has gone into the vexed question of why bread stales. Basically there are two forms of staling.

(1) Gradual evaporation of moisture from the loaf.
(2) Change in the chemical structure of the starch content.

If we briefly examine these, we get an insight into the various ways by which the staling of bread may at least be delayed.

1. It is obviously an advantage to make bread having as high a moisture content as possible by making soft doughs or using an improver which will help to retain more moisture (*see* page 53). Mechanical development of the dough such as the Chorleywood bread process will also allow more water to be used in doughmaking. Having now managed to make bread with a higher moisture content we must take steps to see that it is not rapidly lost. Several factors can play their part here:—

(a) Use of short baking time.
(b) Use of steam in the oven.
(c) Proper cooling.
(d) Either wrapping in waxed paper or cellulose film or keeping covered in a tin or container.

2. The change which occurs in the starch molecule during bread staling is termed "chemical" staling and is the result of starch changing its physical state from a soft gel to a hard solid with a loss of moisture. This change takes place once the temperature falls below 55°C (131°F) and can only be arrested by freezing the bread to −5°C (23°F).

This chemical staling is much more rapid near the freezing point of water and therefore in winter it may present more of a problem. Housewives often make the mistake of thinking that their bread will keep longer if it is placed in a normal refrigerator. Staling will, of course, be more rapid at this temperature. This type of staling can be partially reversed by heating the bread to a point above 55°C (131°F), a fact which is often employed to restore old bread by heating it in an oven.

The above refers to the staling of the crumb of bread. Crust staling refers to that state when the crust turns leathery as it absorbs the moisture from the crumb and occurs much more rapidly.

TYPES OF PROCESSES

The following table summarizes the differences between the various factors which need to be considered in making doughs by these processes.

	Short	*Medium*	*Long*
Time in hours	$\frac{1}{2}$–2	3–6	8–16
Flour	Medium	Medium	Strong
Salt %	1·4–1·8%	1·8–2·0%	2·0–2·5%
Temperature	27–29°C (80–84°F)	24–27°C (76–80°F)	23–24°C (72–76°F)
Consistency	Soft	Average	Tight

Straight Doughs

By this we mean doughs which are made in one stage, i.e. all the ingredients are mixed together in the dough which will be ultimately made into bread.

These may be as follows:

(*a*) *No-time doughs* Doughs which are scaled and moulded immediately they are made, i.e. given no bulk fermentation time. Such doughs are usually made with dough temperatures up to 32°C (90°F) e.g. *Hovis*.

(*b*) *Short fermentation process* Doughs which are given only a few hours bulk fermentation, e.g. 1 or 2 hours. Most bakers to-day make their bread on this type of short process. Not only can fluctuations of temperatures be more closely controlled but yield is not sacrificed.

(*c*) *Medium fermentation process* From 3–6 hours fermentation.

(*d*) *Long fermentation process* These are given six or more hours bulk fermentation. It was usual, before the 1939–45 war, to find bakers making overnight doughs, particularly if they worked almost single-handed. Such bakers were beset with problems when the temperature of the weather fluctuated between the time at which the dough was made and the time at which it was scaled. Bread from dough made under such conditions usually had a very good flavour but was often either over- or under-ripe. Some bakers found an answer to this by the use of a sponge and dough.

Sponge and Dough

In this process a very slack dough is made from most of the water and the yeast but only a proportion of the flour. After the "sponge" as it is now called had fermented for the required time, the rest of the flour and salt, etc. is mixed in it to form the dough. A short process of 2 hours approx. involving a sponge is called a "flying sponge".

For overnight sponges it is usual to leave approx. 6–7% (1 gallon, 4·5, per sack) of water out of the initial sponge and add this at the time of incorporating the rest of the flour. The temperature of the finished dough can thus be adjusted by the temperature of this added water. It is safer, however, to make up long sponge processes with cold water to minimize excessive temperature fluctuations.

The extra work entailed by this two stage process is offset by the above obvious advantage, and the better flavoured bread which results. A saving in the yeast quantity may also be made.

When making sponge and doughs three other points need to be considered.

(1) Since a more thorough ripening takes place a stronger flour may be used.
(2) Considerable softening takes place so that the longer the fermentation, the stiffer it should be made.
(3) The shorter the process the more flour which should be used in the sponge so that a greater bulk can be ripened.

Delayed Salt Method

In this process the salt is omitted at the initial stage of doughmaking but added by dusting into the dough at the knock back stage. It is important that the salt should be free flowing to execute this properly and it might be necessary to sieve it first. Lumps of salt do not easily disperse in a medium like dough and could find their way into the finished loaf.

The advantage of leaving out the salt for up to $\frac{3}{4}$ of the bulk fermentation time is as follows:–

(1) A more thorough ripening of the gluten results in a finer crumb structure.
(2) Crumb stability is improved.
(3) Fuller fermentation results in a better flavour.
(4) The dough behaves better when subjected to automatic plant processes giving better release characteristics yet exhibiting a stickiness which helps to seal the seam of the moulded loaves.

Other Processes

Over the last decade, there has been a number of revolutionary breadmaking processes many of which have now been adopted in this country and abroad by the large producers for the mass production of plant bread. Some of these processes involve a high capital outlay for specialized machinery and the use of chemicals to bring about the conditioning of the dough normally brought about by yeast. Such processes enable doughs to be made on a no-time process which improves yield and provides a more accurate control of the whole process.

The processes in current use rely upon the conditioning of the dough either by–

(*a*) Mechanical dough development, e.g. Chorleywood bread process, or

(*b*) Activated dough development using reducing and oxidizing agents.

(*c*) Delayed ingredients process–with the inclusion of enzyme-active soya or a proprietary improver.

There are modifications of these various processes but in general they all fall within these three categories. Some like (*a*) require high speed mixers whilst others like (*b*) and (*c*) can be made on conventional slow open pan mixers.

As this book is essentially a craft publication, it would be inappropriate for details of these methods to be included, especially as there are many books already in print dealing with these plant processes. In any event, such rapid progress has been made over the last few years in this field that it is possible for new processes such as gas injection to take over in the not too distant future, causing these processes to become obsolete.

However, by whichever process the dough is made, its processing into the many varieties described in the next chapter will be the same.

ROPE

This is a disease of bread caused by certain species of the *Bacillus subtilis*. The spores of this bacteria are resistant to heat and therefore are not always killed in the baking process. If spores of this bacteria have survived and are then given a warm environment in which to multiply, we have the onset of this trouble. It usually occurs in warm weather in bread which is stored in a warm condition. Often bread is stacked closely together as soon as it is baked and this delays the cooling process encouraging the onset of this disease.

The presence of rope manifests itself by liquifying the crumb of a loaf so that when cut and the two halves separated, long strands of material are formed between the two cut surfaces. The sticky crumb is accompanied by a smell of over-ripe fruit reminiscent of pineapples. Although such bread is unpalatable to eat, it is nevertheless harmless.

The following precautions are recommended for the prevention of rope.

(1) Bake thoroughly.

(2) Cool as rapidly as possible and maintain a cool temperature for the storage of bread.

(3) Acidulate the dough by adding one of the following:–

(i) Acetic acid:	98% (Glacial)	1 part diluted with 20 parts water
	80%	1 part diluted with 16 parts water
	30%	1 part diluted with 6 parts water
Spirit Vinegar	12%	1 part diluted with 1 part water
Ordinary Vinegar		Use undiluted

(ii) Acid calcium phosphate (A.C.P.)

(iii) Acid sodium pyrophosphate (cream powder).

(iv) Cream of tartar.

(v) Propionic acid (sodium and calcium propionates).

Quantities To Use Per Kilo ($2\frac{1}{4}$ lb) of Flour

Diluted acetic acid or vinegar	10 g ($\frac{1}{3}$ oz)
Acid calcium phosphate (80% acidity)	6 g ($\frac{1}{5}$ oz)
Acid sodium pyrophosphate	6 g ($\frac{1}{5}$ oz)
Cream of tartar	6 g ($\frac{1}{5}$ oz)
Propionic acid	3 g ($\frac{1}{10}$ oz)

The occurrence of this disease is rare in Britain today because its cause is well understood. If bread is thoroughly baked and stored in a cool place afterwards, this trouble should not occur.

Rope does not usually occur in cake presumably because of its high sugar content although the author has seen an outbreak of rope in a seed cake very lean in sugar made during the rationing period of the last war.

MOULD

Mould is simply a growth of a certain plant-like micro-organism which is propagated by spores. These spores become airborne and so easily contaminate food on which they settle and grow into a colony, sending out delicate threads called mycelia which branch in all directions into the food material. The felt-like mass which results makes food look unpalatable but is not in fact harmful and if removed the rest of the uncontaminated food underneath can be eaten.

Mould generally flourishes in a damp environment and especially in acid conditions. Bread kept in a damp store for example will be much more prone to mould attacks than bread kept in a dry one. Sliced and wrapped bread is more susceptible to mould growth because of the high humidity which builds up inside the wrapper and the extra number of crumb surfaces which are exposed on the slices.

To reduce the incidence of mould the following recommendations should be observed:–

(1) Strict bakery hygiene; removal of all areas of stale food etc. which can harbour the spores. Regular cleaning of equipment such as the blades on slicing machines. Use of a vacuum cleaner for floors etc. to prevent distribution of dust which will cause spores to become airborne.

(2) Prevention of airborne dust from outside with proper air filtration if possible.

(3) Prevention of dampness. Use of anti-mould paints in stores etc.

Anti-Mould Agents

Certain substances called anti-mould agents are now permitted to be used in foodstuffs within strict limits.

Bread

Add at doughmaking stage.

Legal limits:

Propionic acid	Up to 0·3% of flour weight
Calcium propionate	Up to 0·377% of flour weight

The following quantities are usually recommended:

Propionic acid	0·1% of flour weight
Calcium propionate	0·2% of flour weight

The use of the above agents will impart a characteristic taste which, although not unpleasant, may be detected by some customers. It will also retard fermentation and the yeast quantity needs to be increased or the bulk fermentation time extended.

Flour Confectionery

Add at the mixing stage.

Legal limits:

Sorbic acid — Up to ·1% of the weight of the finished product.
Potassium sorbate — Up to 0·134% of the weight of the finished product.

The following levels based upon the finished product are usually suitable:

Madeira, sponge, genoese, etc.:	0·05 to 0·13%
Lightly fruited cakes:	0·04 to 0·08%
Heavily fruited cakes, jams, fruit fillings:	0·01 to 0·04%.

References

Flour Milling & Baking Research Association, Chorleywood, Herts.
Up-to-date Breadmaking by W. J. Fance & B. H. Wagg.

REFRIGERATION

Refrigeration is a great asset to the baker and confectioner, not only to keep perishable raw materials, but also made-up products for long periods of time. Products can be stored either unbaked or baked, to suit the type of trade in question. By deep freezing we can make products in larger batches thereby eliminating production peaks.

Deep Freezing

By this we mean the rapid reduction of temperature below the freezing point of water.

Normal deep freeze temperatures for bakery goods is in the region of $-23°C$ ($-10°F$).

Modern deep freeze units designed for bakery goods are either (a) cabinet size containing racks into which full size baking sheets can be accommodated, or (b) a walk-in room into which racks may be wheeled.

Whichever type of deep freeze installation is used, there are a number of important considerations to be observed.

(1) Because the rate of freezing is dependant upon the weight of the materials being loaded and their temperature, it is important to reduce the temperature of goods to be deep frozen to that of the surrounding air.

(2) The volume of goods to be frozen at any one time must be carefully considered in relation to the capacity of the deep freeze. Manufacturers give guidelines regarding this, and it is wise to follow them. Goods should be placed into the deep freeze in small batches and not filled in one operation.

(3) The size of the individual items will make a difference between the rate at which the deep freeze is loaded. Large items take longer to freeze than small ones, and this must be considered.

The following examples show how long it takes for the product to be reduced from room temperature to 17·7°C (0°F)

Small goods	30–90 *mins.*
$\frac{1}{2}$ *kg*	4 *hours.*
1 *kg*	6 *hours.*

(4) Loss of flavour can be a problem with goods stored in deep freezers, especially if the storage is prolonged. There can also be transfer of aroma from one product to another. For this reason strong smelling substances such as fish etc. should never be stored with bakery goods. Wrapping of the goods is one method by which flavour may be retained, and this will also prevent its transfer.

(5) Products stored in a deep freeze must be removed in strict rotation, and this can be effected by code marking or date stamping.

Rate of Freezing

Rapid freezing of baked goods is essential for two reasons:

(*a*) Slow freezing of water results in large ice crystals being formed which destroy the texture of the material in which it is contained and may result in deposits of moisture forming during defrosting. For example, if a strawberry is slowly frozen, on defrosting we would have a runny pulp, the whole of its structure being destroyed. Rapid freezing on the other hand creates minute crystals of ice that do no damage to the texture, and so preserves its characteristics after defrosting.

(*b*) The second reason relates to staling. When baked products containing flour, especially bread, are allowed to cool to approx. 4·4°C (40°F) the staling rate is rapidly accelerated. It is therefore important to get goods below this temperature range as quickly as possible.

Nitrogen Freezing

Nitrogen freezing of baked goods is strongly recommended. In this plant, liquid nitrogen at $-200°C$ is allowed to escape into a chamber through which baked goods are passing on a conveyor belt. Almost instantaneous freezing of the goods results. Soft fruits such as fresh strawberries, individually frozen by passing them through liquid nitrogen, are almost as good as they were prior to being frozen.

Humidity

A relative humidity of about 80% is necessary if goods are not to become dehydrated before freezing. The wrapping of goods before they are frozen is one way of ensuring that goods maintain their moisture.

FAULTS IN BREAD & FERMENTED GOODS

FAULTS \ CAUSES	Under-ripe dough	Over-ripe dough	Tight dough	Slack dough	Uneven distribution of oven heat	Underproof	Overproof	Baking time too long	Baking temperature too low	Baking temperature too high	Insufficient oven steam
Poor volume & oven spring	√		√			√				√	
Flying top	√				√	√				√	√
Cauliflower top		√		√			√				
Flat appearance		√		√			√		√		
Pale crust colour and lack of bloom		√							√		
Excessively brown crust colour	√									√	√
Lack of bloom		√		√			√		√		
No oven spring		√		√			√		√		
Thick crust	√		√					√		√	√
Elongated holes in texture	√					√					
Large holes in texture		√		√		√	√				
Uneven texture	√	√	√	√		√	√				
Close texture	√		√			√					
Lack of crumb stability		√	√				√				
Poor crumb colour		√		√		√	√				
Poor crumb softness	√		√			√		√	√		√
Poor shelf-life		√	√			√	√	√	√		√
Poor flavour		√									
Poor shape	√	√	√	√	√	√	√				√
Cores in crumb				√							
Sticky crumb		√		√							
Blisters on crust	√						√				
Leathery crust				√							
Sourness		√					√				
Mould				√							
Rope									√		
Holes under the top crust			√								
Break-down of crust surface		√									
Wrinkled crust							√				

Excessive oven steam	Flour deficient in maltose	Flour with high maltose	Use of malt or sugar in recipe	Flour used was too strong	Flour used was too weak	Poor moulding technique	Lack of salt	Excessive salt	Dough temperature too high	Bread stored in damp conditions	Bread stored in warm conditions	Poor mixing (Doughmaking)	Leaving dough exposed	Use of too much fat	Insufficient fat, etc.	Use of excessive mineral improver.	Excessive steam in prover	Lack of recovery time	Too much top oven heat	Wrapping too warm
	√				√			√												
									√											
				√			√		√											
√					√			√												
√	√						√													
		√	√					√												
√	√						√		√											
√					√				√											
				√					√							√				
				√																
					√	√			√											
						√		√												
	√							√								√				
					√	√	√		√					√						
	√				√	√	√		√											
	√			√				√							√	√				
	√			√			√								√	√				
	√						√													
				√		√														
						√						√	√							
		√	√																	
																	√	√		
√										√										
							√		√											
										√										
											√									
																		√	√	
	√																			
																			√	√

TYPE OF GOODS FOR DEEP FREEZING

Not all goods are suitable for deep freezing. When deep frozen goods are defrosted, moisture from the air will become deposited on the cold surface as dew. If this surface is hygroscopic such as a sugar icing, the sugar will first absorb and then become dissolved in the moisture, and run off the goods.

Items which can be successfully frozen are as follows:

Baked Goods

(1) Bread and other fermented goods.
(2) Cakes, plain and fruited.
(3) Chemically aerated goods.
(4) Short and sweet pastries.
(5) Puff pastries.
(6) Almond goods.

Unbaked Goods

(1) Rich buns and rolls.
(2) Cake batters.
(3) Cake fillings including whipped fresh cream.
(4) Chemically aerated goods.
(5) Short and sweet pastries.
(6) Puff pastry.

Defrosting

Rapid defrosting is just as important as rapid deep freezing, especially with baked products such as bread in order to reduce the time that it is passing through the critical staling temperature range. Unbaked goods are defrosted by merely placing the goods in the warm bakery, but it may be speeded by placing in a warm prover with some steam to prevent drying and skinning, before being baked off.

Retardation

This is the technique of slowing down the fermentation of doughs containing yeast, so that they may be left in a retarding cabinet for a period of hours after making, before being baked off.

The temperature employed for this process is 0° to 3·3°C (32° to 38°F).

There are two methods by which fermented doughs may be retarded.

(*a*) *In bulk* The dough is given between half and two thirds of its bulk fermentation time. It is then divided into conveniently sized pieces, flattened, wrapped in polythene and put into the retarder. When required, it is removed, allowed to warm up to room temperature at approx. 21°C (70°F) and then used in the usual way.

(*b*) *In Units* Dough is given half to two thirds its bulk fermentation and made into the desired individual products. These are placed upon baking sheets and immediately transferred to the retarder cabinet where they are left until required. They are then removed, allowed to warm up to room temperature and, if necessary, proved to the right degree before baking off in the usual way.

Retarding cabinets are so constructed as to maintain a relative humidity of approx. 80%, so as to prevent skin formation on the dough. However, this can also be prevented by the use of polythene sheeting particularly for covering goods once they leave the retarder.

Goods containing a high proportion of enriching agents may be retarded successfully up to 72 hours. Leaner goods require less time, up to 48 hours.

4. *White Breads*

METHOD OF MAKING BREAD BY HAND

(1) Weigh all ingredients except water.
(2) Sieve flour onto a clean table or bench.
(3) Place the yeast, salt, fat or oil into separate bowls.
(4) Powders like milk powder and/or improvers can be sieved and mixed with the flour.
(5) Make a bay or well in the flour on the bench.
(6) Weigh the water at the required temperature (*see* page 49). Pour a little onto the yeast and some onto the salt. (The best way to weigh the water is to pour it into a container which has been previously counter-balanced with another (on the scales).)
(7) Pour the rest of the water into the bay, add the dissolved salt, dispersed yeast and oil or fat. If the latter is used, rub it into part of the flour. For alternative method of adding salt see page 58.
(8) Gradually make a thin batter by mixing flour with the liquor from the inside of the bay.
(9) With the hands spreadeagled, lift the rest of the flour through the batter until every part is wetted.
(10) Squeeze and rub the mixture onto the bench (kneading) until a smooth dough has been made free from lumps.
(11) Check the consistency and add more water if necessary (*see* page 52).
(12) Continue to knead the dough for 10 minutes until it is tough and produces a dry skin.
(13) Place and keep the dough covered in a box at a temperature not below that of the finished dough, for its bulk fermentation time (B.F.T.).
(14) When a dough has received about 3/4 of its B.F.T. it may be knocked back, i.e. gas expelled and the dough given a good kneading.
(15) At the end of its B.F.T. the dough is scaled at the appropriate weight and handed up (shaped to a ball).
(16) It is now left covered for a short recovery period of about 10–15 minutes after which it is moulded into its final shape.
(17) The moulded dough shapes are now either left to prove in a covered box or placed in a special cupboard with steam called a prover.
(18) When the dough pieces have risen to the required degree (approximately double its original volume) it is placed into the oven to bake.
(19) After baking, the bread is turned out and placed onto wires to cool properly. Bread should not be left in its tin or placed upon a flat impervious surface, otherwise sweating will occur.

Overnight Sponge and Dough Processes for Recipes on Pages 66 and 67.

The overnight sponge and dough process has certain advantages over the straight dough process, as follows:

(1) Once made, the dough becomes a no-time dough and can be scaled immediately.
(2) The fermentation of the sponge develops flavour which is imparted into the finished bread.

RECIPES USING AN

(*See page* 65

	Sponge White				White Tin				White Crusty			
	kg	*g*	*lb*	*oz*	*kg*	*g*	*lb*	*oz*	*kg*	*g*	*lb*	*oz*
Strong white flour	1	000	2	4	1	000	2	4	1	000	2	4
Wheatmeal		—		—		—		—		—		—
Cold water 15°C (60°F)		500	1	2		530	1	3		485	1	1½
Yeast		10		⅓		35		1¼		35		1¼
Sponge (white)		—		—		750	1	11		375		13½
Sponge (brown)		—		—		—		—		—		—
Salt		—		—		35		1¼		35		1¼
Shortening		—		—		15		½		20		⅔
Improver		—		—		7		¼		7		¼
Separated milk powder		—		—		—		—		—		—
Totals	1	510	3	6⅓	2	372	5	5¼	1	957	4	*6½

*Approximately

(3) Fluctuations of temperature have little effect upon the finished dough because it has no bulk fermentation time. If the sponge has been allowed to ferment at a higher temperature than normal, very cold water can be added at doughmaking to reduce the temperature. This process produces a consistent product, in spite of fluctuations in the weather.

Making the Sponge

(1) Mix the ingredients to a smooth dough and place into a covered container.
(2) Leave for 24 hours at room temperature 18–21°C (65–70°F).

Doughmaking

(1) Make the dough in the usual way.
(2) Scale immediately and hand up.
(3) After approximately 10 minutes recovery, mould, prove and bake according to the variety made.

BREAD AND ROLLS

Yields

In Britain the weight of loaves of bread sold to the public is legally controlled. Since metrication the weight of whole loaves exceeding 300 g in weight may only be made for sale in a net weight of:

(1) 14 ounces (397 g) or a multiple of 14 ounces *or*
(2) 400 g (14·1 ounces) or a multiple of 400 g.

There is an exemption to the above if the loaves are of 10 ounces or 300 g (10·6 ounces) or less.

The reader needs to be aware that the yields given of the recipe will not always divide into an even number of loaves of the required weight. To take account of baking losses we need to scale the dough as follows:

For a 14 ounce loaf – scale at 16 ounces.

For a 400 g loaf – scale at 450 g.

It so happens that if we take a sack recipe, i.e. 280 lb and translate this into grams we produce a dough the weight of which is roughly equivalent to 450 g.

OVERNIGHT SPONGE
for details)

White Danish				Milk				Sponge (brown)				Wheatmeal Tin			
kg	*g*	*lb*	*oz*	*kg*	*g*	*lb*	*oz*	*kg*	*g*	*lb*	*oz*	*kg*	*g*	*lb*	*oz*
1	000	2	4	1	000	2	4		—		—		—		—
	—		—		–		–	1	000	2	4	1	000	2	4
	585	1	5		530	1	3		640	1	7		625	1	$6\frac{1}{2}$
	30		1		35		$1\frac{1}{4}$		7		$\frac{1}{4}$		35		$1\frac{1}{4}$
	305		11		595	1	$5\frac{1}{2}$		—		—		—		—
	—		—		—		—		—		—		375		$13\frac{1}{2}$
	20		$\frac{2}{3}$		30		1		—		—		35		$1\frac{1}{4}$
	20		$\frac{2}{3}$		30		1		—		—		25		$\frac{3}{4}$
	7		$\frac{1}{4}$		7		$\frac{1}{4}$		—		—		—		—
	10		$\frac{1}{3}$		60		$2\frac{1}{4}$		—		—		—		—
1	977	4	7*	2	287	5	$2\frac{1}{4}$	1	647	3	$11\frac{1}{4}$	2	095	4	$11\frac{1}{4}$

The following basic tin bread recipe will illustrate this point:

Tin Bread

	kg	*g*	*lb*	*oz*	*Parts by Weight*
Strong flour	1	000	2	4	280
Water		570	1	$4\frac{1}{2}$	160
Salt		20		$\frac{2}{3}$	5
Yeast		20		$\frac{2}{3}$	5
Totals	1	610	3	$9\frac{5}{6}$	450

B.F.T. – 2 hours.
Dough temperature – 76°F.
The parts by weight can be regarded as a sack recipe based upon 280 lb/127 kg of flour and may be useful if calculations or measurement of materials based upon the sack of flour are required.

To summarize:
Basic Tin Bread Recipe

Basic Sack Recipe		*Gram Recipe*
Strong flour	280 lb	280 g
Water	160 lb	160 g
Yeast	5 lb	†5 g
Salt	5 lb	5 g
	450 lb	450 g

The gram recipe for tin bread will therefore theoretically produce one loaf to be scaled at the required 450 g (1 lb).

† For very small quantities this may have to be increased (*see* page 51).

Let us suppose that we require a recipe to produce 10 loaves. Based upon the sack recipe this would read:

	kg	*g*	*lb*	*oz*
Strong flour	2	800	6	4
Water	1	600	3	8
Yeast		50		$1\frac{3}{4}$
Salt		50		$1\frac{3}{4}$
Totals	4	500	9	$15\frac{1}{2}$

However, it is almost impossible to produce a dough to give a specific yield unless it is made under precise conditions with standardized materials.

The following will all have some effect upon yield:

(1) Water absorption properties of the particular flour being used.

(2) Scaling losses due to particles of dough being left either on the work table or mixing bowl.

(3) Fermentation losses caused by the yeast changing the sugar in the dough into carbon dioxide gas. This loss will increase with the length of fermentation given.

(4) Evaporation losses. This can be kept to the minimum if the dough is kept in a humid environment or kept covered from the atmosphere.

With these factors in mind the author felt that although the kilogram bread recipe fails to produce an exact number of loaves of a fixed weight (there are so many variables which affect the weight of the dough made) there is insufficient justification in altering the principle of kilogram recipes which this book has established.

However, as a guide, if the basic kilo recipe is multiplied by 1·5 ($1\frac{1}{2}$) we get a mass of dough sufficient for 7 loaves @ 450 g (1 lb) scaled weight.

Before proceeding with tin bread recipes some notes about bread tins or pans is not inappropriate. These are made from a number of different materials as follows:

Types of bread tins

Black Sheet Iron This has an oxidized surface which has a degree of rust resistance and a high conductivity of heat.

Tin Plate This is steel which has a coating of tin. Before use these have to be washed and placed in an oven at approximately 204°–218°C (400–425°F) for at least 3 hours for the surface to become dulled. (Ensure that this temperature is not exceeded otherwise the tin plating will melt.)

Aluminium The advantage of these is its good heat conductivity, light weight and resistance to corrosion. However, it lacks strength and to minimize this aluminium-coated steel plate is often used. Aluminium pans are more expensive than many other types.

Electro-Cladded Tins These are first made out of 22 gauge steel and then heavily coated with tin electrolytically. The matt surface of such tins needs no seasoning as with other tins and has better release properties.

New developments in bread tins involve welded seams and radial corners so that a more efficient release of the loaf from the tin is possible.

Tins may also be obtained strapped together in combinations of two to six with spaces allowed between tins for correct heat circulation.

Grease can be applied by brush or cloth, or emulsion can be sprayed to the inside of the tin to aid release of the loaf. Also a special process by which silicone resin can be

WHITE BREADS
(See page 70 for details)

	Tin				Scotch Pan				Crusty				Bloomer				Enriched			
	kg	*g*	*lb*	*oz*	*kg*	*g*	*lb*	*oz*	*kg*	*g*	*lb*	*oz*	*kg*	*g*	*lb*	*oz*	*kg*	*g*	*lb*	*oz*
Strong flour	1	000	2	4	1	000	2	4	1	000	2	4	1	000	2	4	1	000	2	4
Water		570	1	$4\frac{1}{2}$		570	1	$4\frac{1}{2}$		530	1	3		530	1	3		585	1	5
Yeast		30		1		30		1		30		1		30		1		30		1
Salt		20		$\frac{2}{3}$		20		$\frac{2}{3}$		20		$\frac{2}{3}$		20		$\frac{2}{3}$		20		$\frac{2}{3}$
Shortening		—		—		10		$\frac{1}{3}$		—		—		20		$\frac{2}{3}$		—		—
Butter		—		—		—		—		—		—		—		—		20		$\frac{2}{3}$
Skimmed milk powder		—		—		20		$\frac{2}{3}$		—		—		20		$\frac{2}{3}$		30		1
Sugar		—		—		5		$\frac{1}{6}$		—		—		5		$\frac{1}{6}$		—		—
Totals	1	620	3	$10\frac{1}{6}$	1	655	3	$11\frac{1}{3}$	1	580	3	$8\frac{2}{3}$	1	625	3	$10\frac{1}{6}$	1	685	3	$12\frac{1}{3}$

Dough temperature 24°C (76°F)
Bulk fermentation 2 hours
Knock back at $1\frac{1}{2}$ hours

applied to the inside of the tin is available by which a large number of bakings can be made without the necessity for greasing.

White Tin Bread

The basic recipe for a straight dough on page 67 may be improved by additives (*see* page 46), *see* pages 66 and 67 for overnight sponge and dough recipes.

(1) Allow the dough to receive its required bulk fermentation.
(2) Scale the dough into pieces of the required weight.
(3) Lightly mould round or hand up and leave to recover for approx. 10 minutes under a cloth or plastic sheet.
(4) During this time warm the bread tins and lightly grease them.
(5) Mould the dough into the appropriate shape and place into the tin with the seam at the base.
(6) Stand in a warm place to prove but keep covered to prevent skinning. Alternatively, place in a prover with a little steam.
(7) When the dough pieces have proved sufficiently (approx. twice their original volume) place into an oven at 246–260°C (475–500°F).
    The time of baking depends not only upon the temperature but also the size of the loaf (*see* page 55).
(8) Once baked the loaf should be removed as soon as possible from its tin and cooled. Stacking hot bread on an impervious surface or together should be avoided. This causes sweating the steam from the bread condensing to form moisture which will become absorbed by the crust and cause damp patches and encourage mould.

Note The white tin bread recipe is also suitable for making into bread rolls which are usually scaled at 55 g (2 oz) each.

TYPES OF TIN BREAD BY SHAPE

Many types of tin bread are identified by shape although they can all be produced from the same basic dough.

Long Tins

For this type of loaf the dough is moulded in a sausage shape and placed into a long narrow tin to prove and subsequently bake.

Such bread can be cut into a larger number of smaller slices than conventional tin bread and may be preferred on this account.

Split Tins (*see* Figure 2)

To achieve this shape the top of the dough is split lengthwise along the centre with a sharp knife just prior to placing it into the oven. During baking the top splits open and this increases the amount of crust area which improves flavour.

Provided the tin is sufficiently large or the dough mass contained of the correct weight, a baking sheet can be placed over, or the tins inverted onto baking tins so that the top becomes flattened giving square slices when cut.

Sandwich Bread

This is the name given to bread baked in a fully enclosed rectangular bread tin. Care is required in the final stages of proving and in baking to ensure that the expansion is not too great and the dough either completely fills the tin or in extreme cases forces off the lid with which the bread tin is fitted.

Figure 2. Plain tin (*left*) Split tin (*right*)

Figure 3. Sandwich: tin (*left*) loaf (*right*)

Although sandwich bread is usually baked in special tins fitted with lids, it may also be baked in tins inverted on a baking sheet.

Because of the heat required to penetrate the extra metal of the tin, sandwich bread requires a longer baking time – usually 5–10 minutes over open tin bread.

Multipiece Loaves

Here the bread is composed of several pieces of dough placed either concertina fashion in the tin or two long pieces twisted together. The main advantage of this technique is to produce bread which has a finer and more stable texture which is more suitable for buttering purposes.

Pan Coburg

For this shape the dough is moulded round and placed into a round pan.

Just prior to baking, the top is cut $\frac{1}{2}$ inch deep in the form of a cross using a sharp knife. Alternatively, the top may merely be docked or floured and docked.

Scotch Pan

Originally this bread was baked in units of four or five loaves in a long tin, the ends of the dough pieces being greased to enable them to be easily divided. This produced a number of loaves devoid of crust at each end. Today a considerable amount of such bread is made in large plants where it is possible to batch the bread so that it is baked mainly with only a top and bottom crust. Such bread usually contains more enrichment in the form of fat etc. and salt which is a traditional feature of all Scottish bread.

Pistons (or Roundies) (*see* Figure 4 below)

This bread is baked in a long fluted tin. It is in two halves, one being hinged on the other so that once the dough piece has been placed in, the lid can be secured before the loaf is proved and baked. This produces a long ribbed cylindrical loaf which can be cut into round slices.

Farmhouse

This shape is very popular. The dough is baked in a shallow pan either rectangular or oval in shape. Usually a slightly enriched dough is used, i.e. Scotch pan, and the fat used is often butter.

Danish

This name is often given to bread baked in farmhouse tins, dusted with flour and then the top split with a knife.

Crusty Bread (*see* page 69 for straight dough recipes)
(*see* pages 66 and 67 for overnight sponge and dough recipe)

This is bread completely covered by crust and baked on the sole of the oven. Because there is no pan to maintain the dough in a definite shape during baking, the dough required for crusty bread has to be tighter, which means less water content. The unbaked loaves are usually proved on boards covered with a dusting of rice cones to prevent sticking.

TYPES OF CRUSTY LOAVES BY SHAPE

Cottage

This shape is dying out in Britain because it is becoming uneconomic to make, being extravagant in terms of baking losses and labour. Moreover the shape is not convenient to be sliced into sandwiches. Its main virtue is in its crustiness and some people are still prepared to pay extra for this type of bread because of this feature.

It is made as follows:

(1) After scaling, the dough is divided into two pieces, one being twice as large as the other.
(2) Each piece is moulded round and left to recover for 10 minutes.
(3) After recovery, the pieces are remoulded round, flattened, the small piece placed upon the larger piece and flattened.
(4) The thumb of one hand and two fingers of the other hand are pressed into the centre of the top to form a hole which is then enlarged by pulling the hands apart. This is known as *bashing*. Unless a sufficiently large hole is made in the centre of the top piece of dough by these means, a hole is likely to form in the centre of the loaf during baking.
(5) After proving under cover, the cottage loaves should be individually placed onto the sole of an oven in which there should be some steam. Baking time is approximately 40 minutes at a temperature of 246°C (475°F).

These loaves can have the sides split with a sharp knife just prior to baking if desired (knotching). Even baking is essential for the shape of this loaf not to suffer. Uneven baking results in the top either tilting over or coming off altogether.

Figure 5. Oven bottom cottage (*left*) Oven bottom coburg (*right*)

Coburg (*see* Figure 5)

These are made in exactly the same way as the pan coburg, but baked individually on the sole of the oven. Baking times and temperatures are the same as for cottage loaves.

Bloomer* (*see* Figure 6 below, bottom right)

This type of bread originated in London, but is now popular over most of the country because of yielding a good number of slices of reasonable size and with plenty of crust. It is made as follows:

(1) The pieces are handed up and left for 10–15 minutes to recover.
(2) They are then moulded into a long loaf approx. 38 cm (15 in) for 800 g (28 oz) loaf and left to prove. Usually this is done in wooden boxes with sides sufficiently high so that they can be stacked upon each other. This eliminates the need to cover them with a cloth to prevent skinning with the exception of the top box.
(3) After the required proving period, the dough pieces are brushed over with water or more usually a thin starch paste. This is made by first mixing some flour with cold water to form a paste free of lumps. Boiling water is then poured on and the mixture whisked until it thickens. On cooling this thick paste may be diluted to a thinner consistency. The use of this paste will improve the glaze of this loaf after baking.
(4) Approx. fifteen cuts are now made diagonally across the top of the dough with a sharp knife which should be held at an angle, of approximately 45°. If desired, fancy cutting can be made by cutting a series of shorter cuts diagonally along each side of the loaf.
(5) The proved loaves are now placed individually onto the sole of an oven which is filled with steam. Baking times and temperatures are the same as for coburgs and cottage loaves.

Belgian*

This name is often given to a shorter version of a bloomer with approx. five diagonal leaf cuts. Same baking times and temperatures.

Danish*

The loaf more usually known by this name is made from a bloomer dough, moulded to approx. 20 cm (8 in) in length, dusted with flour and given one cut lengthwise with the knife held at an angle. This causes one side to open like a leaf. A slightly cooler oven is recommended, approx. 238°C (460°F).

Figure 6

Tiger Skin Batons* (*see* Figure 6, bottom left)

An interesting decorative surface can be applied to the bloomer shape by use of a paste made as follows:

	kg	*g*	*lb*	*oz*	*Yield*
Water	1	000	2	4	Sufficient for approx. 40 × 800 g (28 oz) loaves.
Ground rice	1	000	2	4	
Yeast		35		$1\frac{1}{4}$	
Sugar		35		$1\frac{1}{4}$	
Oil		70		$2\frac{1}{2}$	
Totals	2	140	4	13	

(1) Mix to a paste with water at 38°C (100°F).
(2) Allow the paste to ferment for 30 minutes.
(3) Stir to eliminate the gas.
(4) Brush the paste liberally onto the loaf prior to baking.

Note The dough pieces may be weighed slightly less to compensate for the weight of the paste being used.

Dutch Scissor*

This bread is shaped like the bloomer but the top is cut with vertical incisions with scissors down each side of the loaf similar to a zip fastener. The sides are cut alternatively just prior to setting in the oven.

**Note* All these types may be made from the overnight sponge and dough of the Danish recipe given on pages 66 and 67.

Other Varieties

The flavour of bread can be varied by adding maw, poppy or caraway seeds either to the bread itself, or as a top dressing. Such bread is very popular in some quarters, particularly Jewish.

Also plaits can be made from these basic doughs or small plaits laid onto boat-shaped pieces. In Jewish circles these are known as *Cholla* although usually a more enriched milk dough is used for such shapes (*see* bun loaves).

MILK BREAD

See page 76 for straight dough recipes.
See pages 66 and 67 for overnight sponge and dough recipes.

In Britain the term *milk bread* can only be applied to bread for sale in which the whole of the liquor used is milk. Legally it must contain not less than 6% of whole milk solids (calculated by weight on the dry matter of bread).

Bread containing the minimum amount of skimmed milk solids may be described as *milk bread* on condition that this term is qualified by such words as "containing milk solids" or "containing separated milk solids" (*see* page 22).

The addition of milk in breadmaking has the following effects:

(1) When full-cream milk powder or fresh milk is used the cream has the same effect as added fats, helping to shorten the crumb and soften the crust and so assist in retarding staling.

MILK BREADS

(See page 75 for details)

	Milk 1				Milk 2				Milk 3				Skimmed milk			
	kg	*g*	*lb*	*oz*	*kg*	*g*	*lb*	*oz*	*kg*	*g*	*lb*	*oz*	*kg*	*g*	*lb*	*oz*
Strong flour	1	000	2	4	1	000	2	4	1	000	2	4	1	000	2	4
Water		585	1	5		585	1	5		70	—	$2\frac{1}{2}$		585	1	5
Milk (3% fat)		—		—		—		—		500	1	2		—		—
Butter		—		—		20		$\frac{2}{3}$		—		—		—		—
Shortening		—		—		—		—		—		—		60		$2\frac{1}{4}$
Whole cream milk powder		60		$2\frac{1}{4}$		—		—		—		—		—		—
Skimmed milk powder		—		—		50		$1\frac{2}{3}$		—		—		60		$2\frac{1}{2}$
Salt		20		$\frac{2}{3}$		20		$\frac{2}{3}$		20		$\frac{2}{3}$		20		$\frac{2}{3}$
Yeast		30		1		30		1		30		1		30		1
Totals	1	695	3	13*	1	705	3	13	1	620	3	$10\frac{1}{6}$	1	755	3	$15\frac{1}{2}$*

*Approximately

Dough temperature 24°C (76°F)
Bulk fermentation 2 hours
Knock back at $1\frac{1}{2}$ hours

(2) The natural sugar of milk, lactose, is unfermented by yeast and so confers an added sweetness upon the finished bread. Also colour and bloom are imparted making it necessary for milk bread to be baked at a lower temperature.
(3) The protein of milk, casein, has a binding and therefore a depressant effect upon the volume of the loaf. To counter this effect extra water is required and additional fat is recommended.
(4) Milk greatly increases the nutritional value of the bread since it contains all the protein and mineral matter necessary for life and the Vitamins A, D and E.

There is no doubt that better bread can be made from recipes containing only 50% milk as in the enriched recipe on page 69 but such bread cannot be sold as *milk bread* and would have to be given some other name such as *fancy tea bread.*

Usually milk bread is sold in units of 300 g (10 oz) or less and so its weight is not legally controlled.

All the shapes described can be made from any of the recipes listed including the enriched recipe No. 5 on page 69.

All the following shapes are baked in an oven at 226°C (440°F) for approx. 30–35 minutes:

Shape No 1 (Undertin) (*see* Figure 7)

(1) Scale at 300 g (10 oz) and mould to fit small rectangular or oval bread tins.
(2) Place the moulded piece onto a lightly greased baking tray and brush over with a wash made from egg and milk.
(3) With a sharp knife cut the top in various decorative patterns and then cover with the bread tin.
(4) When sufficiently proved, bake.

To judge the correct proof leave one loaf upside down in its tin, turning it back when it is ready for baking.

Figure 7. Milk bread baked under tins

Figure 8. Fancy Milk bread: scroll (*top left*); shape no. 3 (*bottom left*); sprial (*top right*); crown (*bottom right*)

Shape No. 2 (Crown) (*see* Figure 8)

(1) Divide the scaled dough piece into seven, six being the same size, whilst the seventh is slightly larger.
(2) Mould round, and arrange in a lightly greased small cottage pan with the larger piece in the centre.
(3) Wash with a mixture of egg and milk and after proof, place into the oven to bake.

Shape No. 3 (*see* Figure 8)

(1) Mould the scaled dough piece round and place into small lightly greased cottage pans.
(2) Wash with egg and milk and with a sharp knife cut the top in the form of diamonds.
(3) Prove and bake.

Shape No. 4 (Scroll) (*see* Figure 8)

(1) Pin out the scaled dough piece into a rectangle and fold into three to form a long flat strip.
(2) This is rolled up swiss roll fashion half way, turned over and repeated from the other side, so that an *S* is formed when turned on its side.
(3) This is then placed into a lightly greased oval bread tin and washed with egg and milk.
(4) Prove and bake.

Shape No. 5 (Spiral) (*see* Figure 8)

(1) Mould the scaled dough piece into a rope.
(2) Roll up to form a spiral.
(3) Place into a greased small cottage pan and egg wash.
(4) Prove and bake.

MILK ROLLS

Figure 9.

Any of the milk bread recipes can be used to make milk rolls, but it is usual to use the enriched white bread recipe, reducing the water by 30 g (1 oz) to make it stiffer.

The following varieties may also be made from the special enriched dough on page 84 for plaited shapes.

All these shapes are made from dough pieces scaled at 55 g (2 oz) washed with a mixture of egg and milk and baked at 226°C (440°F) (*see* Figure 10 below).

1. Knot (*see* Figure. 10(7))

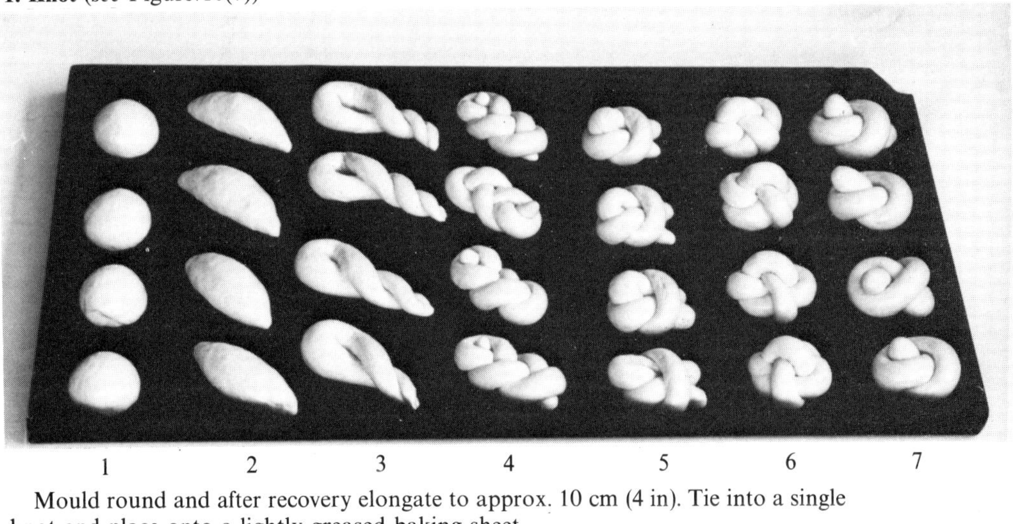

1 2 3 4 5 6 7

Mould round and after recovery elongate to approx. 10 cm (4 in). Tie into a single knot and place onto a lightly greased baking sheet.

2. Winkle

Mould round and after recovery elongate to approx. 15 cm (6 in). Place the piece onto a lightly greased baking sheet in the form of a spiral.

3. Double Knot

Proceed as for 2, but make into a double knot.

4. Rosette (*see* Figure 10(6))

Proceed as for 2. Start by making a loose knot with a long tail, but thread the tail through and tuck under.

5. Button (*see* Figure 10(5))

Proceed as for 2. Make a lover's knot and thread the tail through to form this shape.

6. Layered

Pin out the piece of dough to approx. 13 cm (5 in) diameter and fold into four.

7. 4-Strand Plait

Divide the piece into two and make two long strands. Form the strands in a cross and plait (*see* page 86).

8. 3-Strand Plait

Divide the piece into three and plait into a 3-strand plait (*see* page 85).

9. Twist (*see* Figure 10(3))

Proceed as for 2. Fold in two and twist the two ends together.

10. 1-Strand Plait (*see* Figure 10(4))

Proceed as for 2 and make into a one-strand plait (*see* page 85).

11. Twin

Mould round and after recovery mould into a dumb-bell shape. Roll up each end as in making Italian rolls (Vienna).

12. Cannon

Make a cannon shape from the previous variety (*see* Vienna Rolls).

13. Finger

Mould round and then roll to a finger shape.

14. Creased

Mould round and after recovery press a small rolling pin into the centre.

15. Baton (*see* Figure 10(2))

Mould round and then into a baton shape.

16. Round (*see* Figure 10(1))

Mould up round.

5. *Vienna Bread, Rolls and Plaits*

VIENNA AND FRENCH BREAD

	Vienna				French			
	kg	*g*	*lb*	*oz*	*kg*	*g*	*lb*	*oz*
Strong flour	1	000	2	4	1	000	2	4
Water		585	1	5		585	1	5
Yeast		55		2		55		2
Salt		25		$\frac{3}{4}$		15		$\frac{1}{2}$
Shortening		30		1		30		1
Skimmed milk powder		30		1		—		—
Sugar		5		$\frac{1}{4}$		5		$\frac{1}{4}$
Totals	1	730	3	14	1	690	3	$12\frac{3}{4}$

Dough temperature 21°C (70°F) (the use of ice may be necessary in warm weather).
Bulk fermentation 2 hours.
Knock back every 30 mins.

VIENNA BREAD AND ROLLS

Good Vienna bread should have the following characteristics:

(*a*) Thin, crisp biscuit-like glazed crust. This is achieved by having a cool dough and baking in steam.

(*b*) Open crumb structure with large holes. This is obtained by repeated knocking back, a well fermented dough.

Figure 11. Vienna loaf

VIENNA BREAD

(1) Make up a cool dough 21°C (70°F) in the conventional manner giving it a thorough mixing. If possible, the flour used should be stronger than for other types of bread. Canadian flour or a proportion of Canadian flour added to the normal flour blend is recommended.
(2) Cover the dough and set aside in a warm place.
(3) Knock back every 30 minutes. This will help to develop the open crumb characteristic of this type of bread.
(4) Scale at 300 g (10 oz), hand up lightly and cover.
(5) After 10 minutes recovery, mould to a baton shape. This is done by first flattening the dough piece and then rolling up swiss-roll fashion, applying pressure with both hands to form the shape. Do not make the ends pointed as these will bake quicker and be unsightly.
(6) The moulded dough pieces are now proved either upside down on cloths or on a tray dusted with rice cones. In the former method a long cloth should be used and folded between each loaf (*see* Figure 12). This method eliminates the need to cover the dough pieces and allows the base of each loaf to skin, so facilitating its transfer to the sole of the oven for baking. If the loaves are proved on trays dusted with rice cones, they must be covered to prevent skinning. Proving time should be approx. 25 minutes.
(7) When almost fully proved the tops of the loaves should receive approx. five cuts with a sharp knife about five minutes prior to baking.
(8) The loaves are placed into an oven full of wet steam at a temperature of 226°C (440°F).
(9) Ten minutes after the loaves have been placed into the oven, all the steam is removed and the bread allowed to finish baking in dry heat for a further 15–20 minutes.

Note: Vienna bread and rolls are meant to be consumed as soon as possible after manufacture in order to appreciate the crisp crust. As the bread stales, the crust goes leathery and becomes unappetizing.

Figure 12. Vienna rolls being proved upside down on cloths

VIENNA ROLLS

1 2 3 4

5 6 7 8

Figure 13. Varieties of Vienna rolls

For these, the same dough is used but the pieces are scaled at 55 g (2 oz) each. Such rolls are fashioned in many decorative shapes and should be served fresh at dinner. The dough is made, proved and baked in the same way as for Vienna bread.

A variety of shapes for these rolls are now explained. It is recommended that all these rolls are proved up-side-down on cloths.

1. Baton (*see* Figure 13(7))

Lightly mould the dough piece round and after recovery, shape into a small baton. They can be either short or long.

2. Shell (*see* Figure 13(3))

Mould round and after recovery, flatten one side with the heel of the hand. Fold this flattened portion over the rest of the shape. During baking this flap will open to give a shell-like roll.

3. Kaiser (*see* Figure 13(4))

Commercially these may now be made with a special tool (*see* Figure 14) but the hand method is explained as follows:

Mould round and after recovery, pin out to approx. 7 cm ($2\frac{3}{4}$ in) diameter. Five folds are now made, each done by bringing part of the outside edge to the centre over the thumb and applying pressure with the heel of the other hand to seal it. The last fold is tucked under the first so that the whole roll has the appearance of a rosette.

Figure 14. Use of tool to form the Kaiser shape.

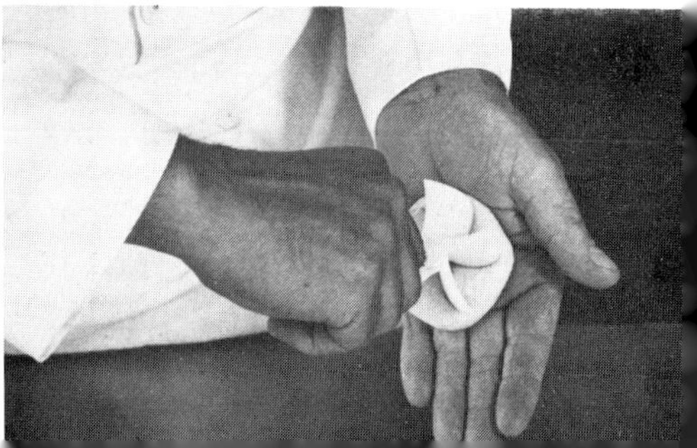

4. Italian 1 (*see* Figure 13(2))

Mould round and then to a dumb-bell shape. Allow to recover and, using a rolling pin, roll out thinly. Roll up like a swiss-roll from each end. Before baking, cut the join with a sharp knife.

5. Italian 2 (*see* Figure 13(5))

After fashioning the roll as previously described, almost sever the roll in half with the side of the hand. Fold one side over to the other.

6. Cannon (*see* Figure 13(6))

Fashion the roll as in (4) but twist one half of the roll over the other to form a shape like a cannon.

7. Crescent (*see* Figure 13(1))

Mould round and after recovery pin out very thinly to a pear shape. Hold the pointed end and tightly roll up swiss roll fashion, keeping tension on the pointed end. When the whole piece has been rolled up it is arranged in the form of a crescent.

8. Creased (*see* Figure 13(8))

Mould round and after recovery press a small rolling pin in the centre.

FRENCH BREAD

(1) Scale the dough into 450 g (1 lb) pieces.
(2) Expel the gas and mould swiss-roll fashion.
(3) Allow to recover and then mould to the required length, e.g. 45 cm (18 in).
(4) Allow to prove, preferably upside-down, on cloths.
(5) Transfer to a wooden slip.
(6) With a sharp knife, cut up to five long slits on the top and immediately place into an oven full of wet steam at a temperature of 226°C (440°F).
(7) After ten minutes baking, remove the steam and finish baking in dry heat for a further 15–20 minutes.

Like Vienna, this bread should be eaten fresh to be fully appreciated. Even after six hours, the crust begins to become leathery and unappetizing.

PLAITED BREADS

The dough required for plaiting needs to be stiffer and have a more stable crumb. This is achieved by using an enriched dough which contains egg as follows:

	kg	*g*	lb	oz
Strong flour	1	000	2	4
Water		470	1	1
Salt		20		$\frac{2}{3}$
Egg		30		1
Shortening		50		$1\frac{2}{3}$
Yeast		30		1
Sugar		20		$\frac{2}{3}$
Totals	1	620	3	10

Dough temperature – 24°C (76°F).
Bulk fermentation – 2 hours.
Knock back at – $1\frac{1}{2}$ hours.

All the plaits are made in a similar manner. The dough piece is first scaled and then divided into the desired number of strands. The weight at which this dough is scaled is a matter of personal preference, but it is recommended that the larger the number of strands which have to be plaited, the greater the scaling weight. These pieces are moulded round and then fashioned into long strands which are tapered at each end (with the exception of the one-strand plait). After plaiting, the shape is washed with a mixture of egg and milk and after proof is baked in an oven at 226°C (440°F). The varieties of plaited shapes are herewith explained and should be read in connection with the diagrams which accompany them.

One Strand

Mould the scaled dough piece round and then make a rope about 3 times as long as the length of the desired plait. Proceed as shown in the illustration (*see* Figure 15 below).

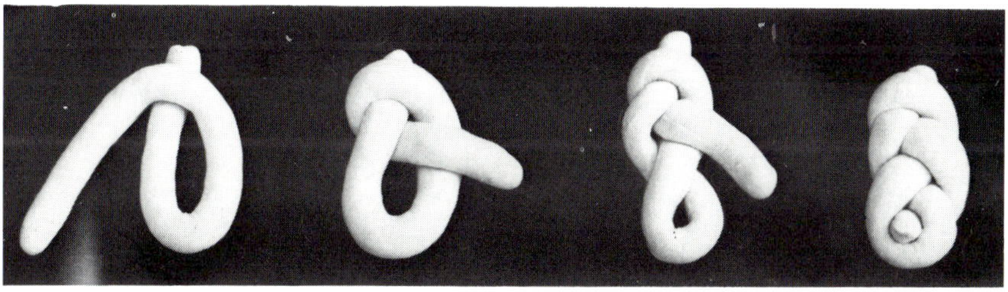

Two Strand

This is not strictly a plait. The dough piece is first fashioned into a rope with tapered ends. It is then folded in half and twisted round each other.

In the following varieties, the dough is first divided into the number of strands required, and each is fashioned into a rope which is bold in the centre and tapered at the ends.

Note The numbers given to the sequences described relate to the position of the strand and not to the strand itself. As the strand is moved during the plaiting operation the remaining strands alter their position relative to each other.

Three Strand (*see* Figure 16)

Lay the three strands together and start plaiting from the centre. When the plait has been done, turn over and complete the other side.

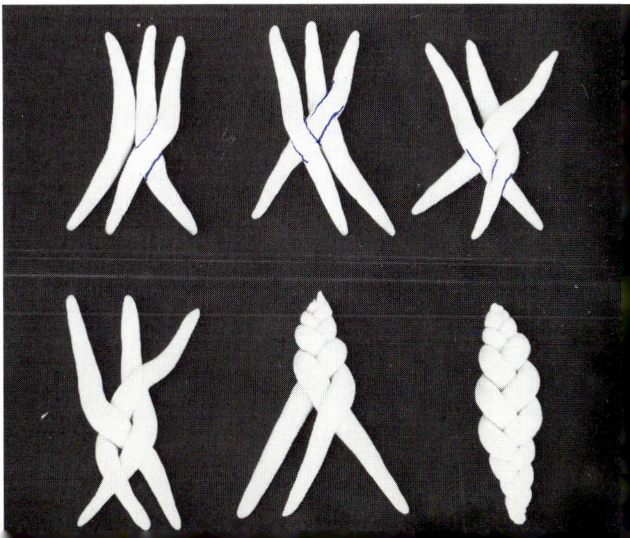

Figure 16. Stages in a 3 strand plait

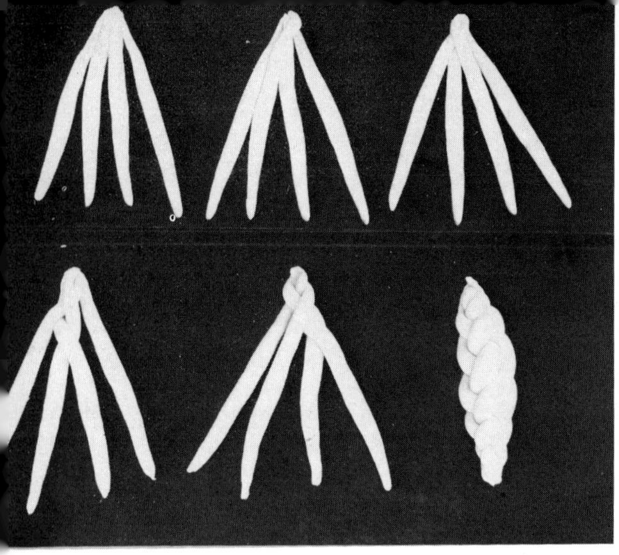

Figure 17 above

Four Strand 1 (*see* Figure 17)

Fasten the four strands at the ends and proceed to plait the strands in the following order, 2 over 3, 4 over 2, 1 over 3, repeat this sequence until the plait is finished.

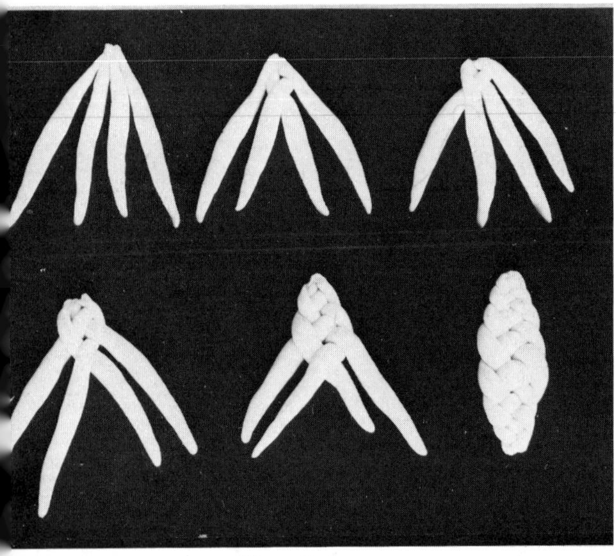

Figure 18 above

Four Strand 2 (*see* Figure 18)

This is plaited in a similar way to the previous plait but is arranged as a flat basket weave.

Figure 19 below

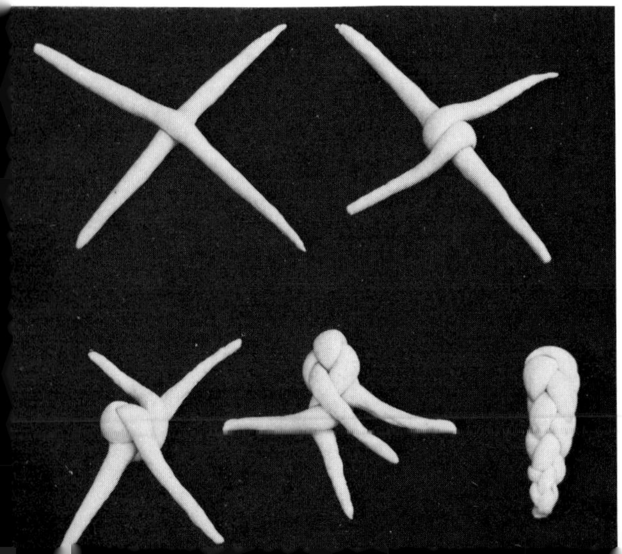

Four Strand 3 (*see* Figure 19)

This is really two long strands. They are arranged in the form of a cross and plaited by passing one strand over the other in turn.

Five Strand (*see* Figure 20)

This is plaited in a similar manner to the four strand but using five strands. The sequence is 2 over 3, 5 over 2, and 1 over 3.

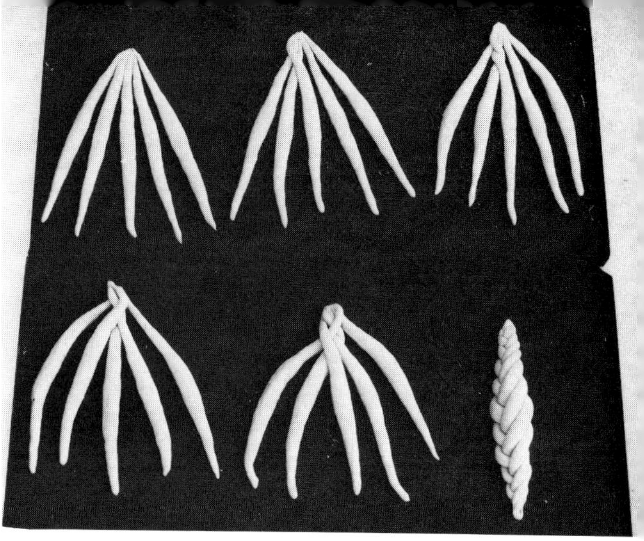

Figure 20

Six Strand (*see* Figure 21)

For these more complicated shapes, the strands need to be longer. In this case the first move is not repeated. This isolated move is 6 over 1. This is followed by the following sequence, 2 over 6, 1 over 3, 5 over 1 and 6 over 4.

Figure 21

Seven Strand

Make seven long strands and place them together on the table so that there are three on one side and four on the other. Start at the centre and bring the strands from the outside to the centre. When one half has been thus plaited turn the piece over and complete by plaiting the remaining half.

Eight Strand (*see* Figures 22 and 23)

Make the eight long strands and fasten together at the ends. This shape also requires an opening isolated move and this is, 8 under 7 and over 1. After this isolated move a series of four moves are repeated until the plait is finished: 2 under 3 and over 8, 1 over 4, 7 under 6 and over 1, 8 over 5.

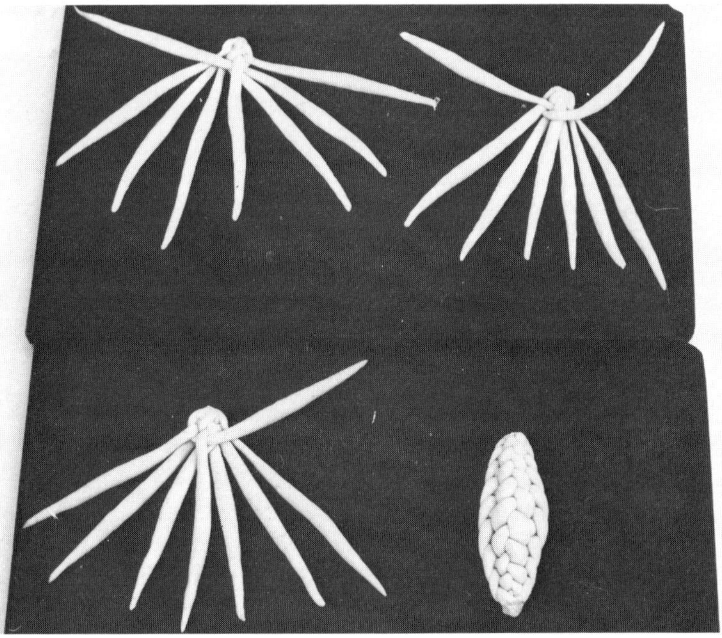

Figure 22 and 23. Stages in plaiting a 8 strand plait

The Winston

This is plaited in the same way as the Four Strand 3, but formed from two lengths consisting of three strands each.

SPECIAL SHAPES

Heart (Centre) (*see* Figure 24)

This is made from two five-strand plaits arranged in the form of a heart. The plaits have to be specially plaited as follows:

(*a*) The strands are shaped with the fat portion at about $\frac{1}{3}$ of its length and with a long tapered tail.

(*b*) One plait is made in the conventional way with the sequences numbered from left to right. The other plait is made conversely with the sequences numbered from right to left.

The two plaits are assembled on a lightly greased baking sheet in the form of the heart, attaching each end. The top attachment can be adorned with a rose also fashioned from the dough. This larger shape requires to be baked at 215°C (420°F).

Figure 24. 8 Strand plait (*left*) plaited heart (*centre*) 3 strand plait (*right*)

Plaited Star (Centre) (*see* Figure 25)

For this ten long tapered strands are made and arranged on a large, lightly greased tray. The strands are now arranged in groups of four and each group plaited as a four strand plait using the basket weave. If the strands are well tapered the result will be an attractively shaped star. For further attraction roses can be fashioned and used to fill the centre. Bake at 215°C (420°F).

Figure 25. 5 strand plait (*left*) plaited star (*top*) 6 strand plait (*bottom*) 4 strand plait (*right*)

6. *Brown and Speciality Breads*

BROWN BREADS

Wholemeal – *See* page 91 for straight dough recipes.
– *See* pages 66 and 67 for overnight sponge and dough recipes

As the name suggests, this bread must be made from wholewheat flour. The Bread and Flour Regulations define wholemeal bread "as being composed of a dough made from the whole of the product derived from the milling of cleaned wheat, yeast and water *without the addition of any other flour* which has been fermented and subsequently baked."

The only genuine wholemeal flour is made by stone milling which is the only milling process which can guarantee that the flour is 100% whole wheat (*see* page 5). These flours contain a high percentage of bran.

The germ of the wheat is also present and the enzymes present help to ripen the dough during fermentation.

These two factors mean that wholemeal bread requires less fermentation than white bread. Fermentation times are therefore either reduced or a smaller amount of yeast is used.

The correct consistency of the dough is difficult to judge because of the water absorption of the bran contained in the flour used. The consistency should not be too stiff.

Because the water absorption of the flour used varies so much, it is impossible to predict with accuracy the correct water absorption and therefore the amounts given must be regarded as approximate.

The bread is made in the same way as for white tin bread (*see* page 65).

Figure 26. Brown bread tin (*left*) Coburg (*right*)

BROWN BREADS

(See page 90 for details)

	Wholemeal Tin				Wholemeal Crusty				Wheatmeal Tin				Wheatmeal Crusty				Cracked Wheat			
	kg	*g*	*lb*	*oz*	*kg*	*g*	*lb*	*oz*	*kg*	*g*	*lb*	*oz*	*kg*	*g*	*lb*	*oz*	*kg*	*g*	*lb*	*oz*
Wholemeal	1	000	2	4	1	000	2	4		—		—		—		—		—		—
Wheatmeal		—		—		—		—	1	000	2	4	1	000	2	4		720	1	10
Whiteflour		—		—		—		—		—		—		—		—		140		5
Cracked wheat		—		—		—		—		—		—		—		—		140		5
Water		720	1	10		695	1	9		640	1	7		610	1	6		640	1	7
Salt		15		$\frac{1}{2}$		15		$\frac{1}{2}$		15		$\frac{1}{2}$		15		$\frac{1}{2}$		20		$\frac{2}{3}$
Lard		20		$\frac{2}{3}$		20		$\frac{2}{3}$		20		$\frac{2}{3}$		15		$\frac{1}{2}$		—		—
Brown sugar		10		$\frac{1}{3}$		10		$\frac{1}{3}$		—		—		—		—		10		$\frac{1}{3}$
Black treacle		—		—		—		—		15		$\frac{1}{2}$		15		$\frac{1}{2}$		—		—
Yeast		20		$\frac{2}{3}$		20		$\frac{2}{3}$		25		$\frac{3}{4}$		25		$\frac{3}{4}$		25		$\frac{3}{4}$
Totals	1	785	4	$0\frac{1}{6}$	1	760	3	$15\frac{1}{6}$	1	715	3	$13\frac{1}{2}$*	1	680	3	$12\frac{1}{4}$	1	695	3	$12\frac{3}{4}$

*Approximately

Dough temperature 24°C (76°F)
Bulk fermentation $1\frac{1}{2}$ hours
Knock back at 1 hour.

Wheatmeal (*see* recipes on page 91).

Because wheatmeal flours contain no germ, they can tolerate longer fermentation times.

By law, wheatmeal bread must contain not less than 0·6% of fibre (calculated on the dry matter of bread). Any additives to white bread may be added (except soya flour) which may be used up to 5 parts to every 100 parts of flour (calculated by weight). Caramel may also be added.

Proceed to make the bread in the same way as for white tin bread.

Cracked Wheat

The dough is made in the same way as wheatmeal. When the dough pieces are moulded, they are rolled or dressed in cracked wheat before being placed into the tin. It is best baked *under-tin* as with milk loaves, *see* page 77.

MALT BREADS

	Light				**Medium**				**Heavy**			
	kg	*g*	*lb*	*oz*	*kg*	*g*	*lb*	*oz*	*kg*	*g*	*lb*	*oz*
Wheatmeal		720	1	10		720	1	10		280		10
Whiteflour		280		10		280		10		720	1	10
Water		555	1	4		530	1	3		500	1	2
Salt		20		$\frac{2}{3}$		20		$\frac{2}{3}$		20		$\frac{2}{3}$
Yeast		35		$1\frac{1}{4}$		35		$1\frac{1}{4}$		55		2
Golden syrup		20		$\frac{2}{3}$		35		$1\frac{1}{4}$		70		$2\frac{1}{2}$
Dry malt extract		35		$1\frac{1}{4}$		55		2		95		$3\frac{1}{2}$
Shortening		20		$\frac{2}{3}$		20		$\frac{2}{3}$		20		$\frac{2}{3}$
Sultanas (optional)		165		6		165		6		165		6
Totals	1	850	4	$2\frac{1}{2}$	1	860	4	$2\frac{5}{6}$	1	925	4	$5\frac{1}{3}$

Dough temperature 24°C (76°F)
Bulk fermentation 1 hour
Knock back at 45 minutes.

Figure 27. Malt bread

There are several proprietary products available which only require water and yeast to be added to produce acceptable malt bread.

The recipes given should make a stiff dough which during fermentation will soften. They should feel quite sticky but this is usual.

All malt breads require slow baking, preferably in a falling temperature. The heavier malted types require an even lower temperature, i.e.:

Lightly Malted – 205°C (400°F) dropping to 193°C (380°F).

Medium Malted – 190°C (375°F) dropping to 182°C (360°F).

Heavily Malted – 177°C (350°F) dropping to 171°C (340°F).

The slow baking necessary with this type of bread demands a long baking time. Lightly malted loaves will take 60 minutes whilst the more heavily malted types up to 1½ hours.

The usual shape is under a long tin so that thc cut slices have as small a surface area as possible.

Many firms make a proprietary malted meal which is usually already salted. All that is required is the addition of yeast and water according to the recommendations given. The following is a typical recipe using such flour.

	kg	*g*	*lb*	*oz*
Proprietary malt meal (salted)	1	000	2	4
Water		445	1	0
Yeast		30		1¼
Totals	1	475	3	5¼

Dough temperature – 24°C (76°F).
Bulk fermentation time – 1 hour.

GERM BREAD

By law wheatgerm bread must contain not less than 10% of added processed wheatgerm (calculated on the dry matter of bread). Up to 5% soya-bean flour and supplementary caramel may also be added. However, this is usually made from flours specially made from wheatmeal with up to 25% added roasted germ. The germ is roasted usually with salt to kill the enzymes which are present. Such flours are made into bread without the addition of salt. Millers of such flours provide recipes and methods for making their own product. The following is typical:

Germ Meal Recipes

	Bread				**Rolls**			
	kg	*g*	*lb*	*oz*	*kg*	*g*	*lb*	*oz*
Germ meal	1	000	2	4	1	000	2	4
Water		695	1	9		595	1	5½
Shortening		7		¼		60		2¼
Yeast		35		1¼		40		1½
Milk powder		—		—		25		¾
Totals	1	737	3	14½	1	720	3	14

Bread

Dough temperature – 31°C (88°F).

Bulk fermentation time – Nil (No-time dough).

(1) Scale at 450 g (1 lb) mould and place into lightly greased tins.
(2) Prove for approx. 30 minutes.
(3) Bake at 232°C (450°F) for approx. 30 minutes.

Rolls

Dough temperature – 27°C (80°F).

Bulk fermentation time – Nil (No-time dough).

(1) After the dough is made, allow to rest for 10 minutes and then scale at 55 g (2 oz) each.
(2) Mould either round or finger shape and place on a lightly greased baking sheet.
(3) Prove for approx. 25 minutes and then bake at 260°C (500°F) for approx. 8 minutes.

GLUTEN BREAD

This is ordinary bread enriched with gluten. This can be obtained by making a well developed unyeasted dough from a strong wheat and then washing out the gluten it contains. The process is time-consuming and most bakers prefer to purchase gluten from a supplier specializing in this product. It may be obtained wet or in the dry form.

The addition of gluten has the same effect as using a stronger flour and in fact if it is used with a weak flour it can be regarded as an improver (*see* pages 00). By its use it is possible to produce large volumed loaves of much less weight than ordinary bread, and since each slice of such bread contains less carbohydrate (because of the extra protein) it is regarded as an aid to slimming. However, legally this cannot be claimed unless it is made perfectly clear that it is part of a diet in which the total intake of calories (or joules) is controlled.

The Bread and Flour Regulations define gluten bread "as that which contains not less than 16% protein calculated on a dry basis", i.e. 16% of the dry matter in the bread must be protein. For high protein bread the limit is raised to 22%.

The baker who elects to make and use his own gluten must know the protein content of the flour he uses, as well as that of the gluten and then calculate the amount he needs to use to fulfil the legal requirements.

Gluten bread or rolls are often termed *starch reduced*, but under the regulations starch reduced rolls must contain 50% gluten and 50% flour.

To summarize, *dry* gluten may be added to bread as follows:

(1) As an improver, particularly to increase the strength of flour.
(2) Gluten bread in which sufficient must be added to increase the protein to 16% based on dry weight.
(3) High protein bread in which sufficient dry gluten must be added to increase the protein to 22% based on dry weight.
(4) Starch reduced rolls 50/50 = gluten/flour.

Method Using Dry Gluten

(1) Blend the flour, dry gluten and milk powder into homogeneous mixture.
(2) Add the fat and mix into the dry ingredients.
(3) Disperse the yeast in a portion of the water.
(4) Add the salt, sugar or other soluble materials to the remaining water.
(5) Add the liquids to the dry ingredients and thoroughly mix to a dough. To ensure a thorough hydration of the gluten the dough should be mixed for approx. 20 minutes by machine.
(6) After bulk fermentation scale at 465 g for a 400 g loaf or 350 g for a 300 g loaf. ($16\frac{1}{2}$ oz for 14 oz or $11\frac{1}{2}$ oz for 10 oz).
(7) Carefully mould and place into warm tins.
(8) Prove well covered and bake at 230°–245°C (450–480°F).

GLUTEN BREAD

(See pages 94 and 96 for details)

	White 1				White 2				Brown 1				Brown 2				High Protein			
	kg	*g*	*lb*	*oz*	*kg*	*g*	*lb*	*oz*	*kg*	*g*	*lb*	*oz*	*kg*	*g*	*lb*	*oz*	*kg*	*g*	*lb*	*oz*
White flour	1	000	2	4	1	000	2	4		220		8		220		8	1	000	2	4
Wheatmeal		—		—		—		—		780	1	12		780	1	12		—		—
Water		625	1	$6\frac{1}{2}$		530	1	3		720	1	10		530	1	3		720	1	10
Dry gluten		50		$1\frac{3}{4}$		—		—		50		$1\frac{3}{4}$		—		—		180		$6\frac{1}{2}$
Wet gluten		—		—		140		5		—		—		140		5		—		—
Milk powder		15		$\frac{1}{2}$		15		$\frac{1}{2}$		15		$\frac{1}{2}$		15		$\frac{1}{2}$		15		$\frac{1}{2}$
Yeast		30		1		30		1		30		1		30		1		30		1
Salt		20		$\frac{2}{3}$		20		$\frac{2}{3}$		25		$\frac{3}{4}$		25		$\frac{3}{4}$		20		$\frac{2}{3}$
Sugar		7		$\frac{1}{4}$		7		$\frac{1}{4}$		—		—		—		—		—		—
Brown sugar		—		—		—		—		7		$\frac{1}{4}$		7		$\frac{1}{4}$		—		—
Shortening		15		$\frac{1}{2}$		15		$\frac{1}{2}$		15		$\frac{1}{2}$		15		$\frac{1}{2}$		15		$\frac{1}{2}$
Caramel colour		—		—		—		—		as req				as req				—		—
Totals	1	762	3	$15\frac{1}{6}$	1	757	3	15*	1	862	4	$2\frac{3}{4}$	1	762	3	15	1	980	4	$7\frac{1}{6}$

*Approximately

Dough temperature 24°C (76°F)
Bulk fermentation 2 hours
Knock back at $1\frac{1}{2}$ hours

Baking times will be approx. 35 minutes for the 400 g (14 oz) loaves and 30 minutes for the 300 g (10 oz) loaves.
Note The consistency of the dough should be slightly slacker than for ordinary white tin bread.

Method Using Wet Gluten

(1) Thoroughly blend the flours and milk powder and rub the fat into the mixture.
(2) Disperse the yeast in some of the water and dissolve the salt, sugar, etc. in the remainder.
(3) Wash the wet gluten thoroughly in water at approximately 49°C (120°F) to bring it into a soft plastic condition.
(4) Mix the dry ingredients with the liquids and once the dough is made, add the wet gluten. Give a thorough mixing by machine for at least 20 minutes to disperse the gluten.
(5) Proceed as previously described for doughs using dry gluten.

Note *Wet* gluten is kept in brine solution. When delivered this should be poured away and refilled with fresh brine solution of $7\frac{1}{2}\%$ strength. This is made by dissolving 75 g salt per litre 1000 g water ($2\frac{3}{4}$ oz per $2\frac{1}{4}$ lb).

Wet gluten may be kept in brine in a refrigerator for a number of days. If after a thorough kneading it does not toughen, then it is not in good condition and should not be used. It is always advisable to use fresh gluten.

CHEESE BREAD

Although no statutory levels are laid down by the regulations, it is recommended that cheese bread should contain at least 10–15% of dry cheese solids, based on flour weight in order to *characterize* the bread as cheese bread.

Powdered cheese made specially for bakers and known as *Bakers' Cheese* is available for the manufacture of this bread. It is quite soluble and is easily distributed throughout the dough.

Grated cheese may be used but apart from the difficulty of preparation, the cheese particles are difficult to disperse in the dough and will produce streaks.

The addition of cheese in the dough will have the following effects:

(1) Cheese has a binding effect which necessitates the addition of more water initially to make a softer dough.
(2) Because the cheese content increases the crust colour, a lower temperature is required as follows:
    White and Wheatmeal – 204°C (400°F)
    Rye – 215°C (420°F)
(3) The lower baking temperature will mean that these loaves will grow considerably in the oven so that they should be proved to only half that of normal tin bread.

The normal scaling weight of such bread is 500 g (16 oz) for a normal 450 g (1 lb) bread tin, or 360 g (13 oz) baked *under* a normal 450 g (1 lb) bread tin. If baked *under-tin* the tray needs to be of double thickness or, alternatively, the loaves may be baked in the normal position with a heavy tray placed on top.

These breads may be baked on the oven bottom provided the dough is made stiffer.

The rye bread variety is usually baked as baton shapes, egg washed and sprinkled with salt and caraway seeds, the latter also being added to the dough if desired.

The dough is made in the usual way, first mixing the powdered cheese with the flour used.

Tomato Bread

This bread is made in the usual way, first adding the tomatoes to the liquid used and bringing it to the correct temperature for making a dough at 24°C (76°F). Pink colour may be added if desired to give the crumb a richer salmon pink colour.

SPECIALITY BREADS

(*See page 96 for details*)

	Cheese Bread												Tomato Bread							
	White				Wheatmeal				Rye				1				2			
	kg	*g*	*lb*	*oz*	*kg*	*g*	*lb*	*oz*	*kg*	*g*	*lb*	*oz*	*kg*	*g*	*lb*	*oz*	*kg*	*g*	*lb*	*oz*
Strong flour	1	000	2	4		—		—		665	1	8	1	000	2	4	1	000	2	4
Wheatmeal		—		—	1	000	2	4		—		—		—		—		—		—
White rye flour		—		—		—		—		335		12		—		—		—		—
Water		640	1	7		695	1	9		695	1	9		335		12		625	1	$6\frac{1}{2}$
Milk powder (skimmed)		30		1		30		1		—		—		25		$\frac{3}{4}$		25		$\frac{3}{4}$
Yeast		55		2		45		$1\frac{1}{2}$		30		1		30		1		30		1
Salt		20		$\frac{2}{3}$		30		1		30		1		20		$\frac{2}{3}$		20		$\frac{3}{4}$
Sugar		15		$\frac{1}{2}$		—		—		—		—		25		$\frac{3}{4}$		25		$\frac{3}{4}$
Shortening		30		1		30		1		—		—		25		$\frac{3}{4}$		25		$\frac{3}{4}$
Bakers cheese		195		7		195		7		250		9		—		—		—		—
Cooked & strained tomatoes		—		—		—		—		—		—		335		12		—		—
Concentrated tomato pulp		—		—		—		—		—		—		—		—		100		$3\frac{3}{4}$
Totals	1	985	4	$7\frac{1}{6}$	2	025	4	$8\frac{1}{2}$	2	005	4	8	1	795	4	16*	1	850	4	$2\frac{1}{6}$

*Approximately

	White	Wheatmeal	Rye	Tomato 1	Tomato 2
Dough temperature 24°C (76°F)					
Bulk fermentation time	2 hours	$1\frac{1}{2}$ hours	3 hours	2 hours	2 hours
Knock back at	$1\frac{1}{2}$ hours	1 hour	2 hours	$1\frac{1}{2}$ hours	$1\frac{1}{2}$ hours

After the bulk fermentation time, it is scaled at 195 g (7 oz) moulded round, flattened and baked under a 450 g (1 lb) coburg tin at 230°C (440°F).

RYE BREAD

	Dark Rye				Light Rye			
	kg	*g*	*lb*	*oz*	*kg*	*g*	*lb*	*oz*
White flour (strong)		500	1	2		805	1	13
Dark rye flour		500	1	2		—		—
Medium rye flour		—		—		195		7
Water		665	1	8		555	1	4
Salt		30		1		25		$\frac{3}{4}$
Black treacle		15		$\frac{1}{2}$		45		$1\frac{1}{2}$
Glucose		—		—		25		$\frac{3}{4}$
Shortening		—		—		—		—
Yeast		25		$\frac{3}{4}$		25		$\frac{3}{4}$
Caraway seeds		25		$\frac{3}{4}$		—		—
Totals	1	760	3	15	1	675	3	$11\frac{3}{4}$

Dough temperatures 24°C (76°F)
Bulk fermentation times 3 hours dark rye, 2 hours light rye
Knock back at 2 hours for dark rye, $1\frac{1}{2}$ hours for light rye

There are four types of rye flour available as follows:

1. White Rye

Milled mainly from the centre of the grain, this product is lighter in colour than the other types. Because it has a low protein content, it lacks strength and is used for light rye breads.

2. Medium Rye

Flour produced from the rye grain after the bran and shell has been removed. It is darker in colour and has greater strength than white rye.

3. Dark Rye

This is milled from the grain after some of the finer starch cells have been removed. It has the darkest colour and is the strongest having up to 16% protein.

4. Rye Meal

This is made by grinding the entire rye berry. It is coarse and dark and used for pumpernickel and rolls.

The use of the cereal rye for breadmaking is very common on the Continent where there is a great variety of rye bread made in a number of different methods, many involving sponge and dough and sour-dough processes. However, the popularity of this bread is growing in Britain and therefore no book dealing with breadmaking is complete without one or two recipes which are reasonably easy to prepare.

Some idea of the differences between rye bread and wheaten bread is as follows:
(1) Since the gluten-forming proteins of rye are unstable and unable to form a good

crumb structure it is usually reinforced with wheaten flour. Proving times are short for this reason.

(2) Doughs are very sticky to handle and have to be made fairly tight.

(3) The dough texture is very close and dense compared with white bread and therefore a cooler baking temperature is required with a longer baking time. Baking temperature and times are as follows:

Dark Rye – 205°C (400°F) for approx. $1\frac{1}{4}$ hours.

Light Rye – 220°C (440°F) for approx. 50 minutes.

The very dark rye bread called Pumpernickel which is made with a high percentage of rye meal, and a sour dough, requires a much slower oven of approx. 160°C (340°F) with a correspondingly extended baking time.

(4) Rye bread has a remarkably long shelf life with the flavour and crumb improving on keeping. The author has eaten black rye bread which was 6 months old and still palatable.

(5) It is recommended that the caraway seeds are crushed with the salt to improve the flavour.

Method

(1) Blend the flours well together and make up the dough in the usual manner.

(2) After bulk fermentation, scale at the statutory weight, allow to recover and then mould as bloomer shapes (*see* page 74).

(3) After proving, wash with a thin starch paste (*see* page 74) and make a number of cuts at right angles, close together. Alternatively, caraway seeds crushed with coarse salt may be sprinkled on.

(4) Bake on the sole of the oven at the appropriate temperature with the oven full of steam.

CHEMICALLY AERATED BREADS

Apart from the facts that the aeration of this type of bread is accomplished by chemicals instead of yeast, there are other differences as follows:

(1) Yeast not only produces carbon dioxide gas in a dough but also imparts a ripening action which makes the dough more extensible and helps to develop a soft moist crumb. Part of this ripening in a dough can be caused by lactic acid fermentation. Baking powder can only produce gas and so in order to achieve some measure of ripening in doughs aerated by this means, buttermilk is often used as the moistening agent. This is the residual liquid from the churning of ripened cream into butter and is practically a solution of lactic acid in water.

(2) The flour needs to be much softer, one of medium strength being required.

(3) Dough consistency is much softer than for conventional yeasted doughs.

(4) Salt quantities are usually less.

(5) Baking must be done at a lower temperature in order to allow the loaves to grow to their maximum volume.

(6) The mixing of the dough should not be too thorough, otherwise toughening will result.

Figure 28. Scotch farl

CHEMICALL
(*See pages 99 ar*

	White soda				Wheaten Soda			
	kg	*g*	*lb*	*oz*	*kg*	*g*	*lb*	*oz*
Soft flour	1	000	2	4		500	1	2
Wheatmeal		—		—		500	1	2
Coarse bran		—		—		—		—
Bicarbonate of soda		20		$\frac{2}{3}$		20		$\frac{2}{3}$
Cream of tartar		30		1		30		1
Baking powder		—		—		—		—
Butter or lard		85		3		85		3
Sugar		10		$\frac{1}{3}$		10		$\frac{1}{3}$
Golden syrup		—		—		—		—
Egg		—		—		—		—
Salt		15		$\frac{1}{2}$		15		$\frac{1}{2}$
Milk		—		—		—		—
Buttermilk		915	2	1		915	2	1
Sultanas		—		—		—		—
Currants		—		—		—		—
Totals	2	075	4	$10\frac{1}{2}$	2	075	4	$10\frac{1}{2}$

Method for Recipes 1–4

(1) Sieve the flour and chemical aerating agents.
(2) Rub the shortening into the flour.
(3) Dissolve the salt and sugar in the buttermilk or milk and add to the rest of the ingredients.
(4) Mix lightly to a soft dough.
(5) Scale at 450 g (1 lb) and mould round.
(6) Flatten slightly and place onto a flat baking sheet.
(7) Wash with a 5% solution of brine [(20 g) ($\frac{2}{3}$ oz) salt in 100 g ($3\frac{1}{2}$ oz) water] and then dust with flour. For the wheaten soda bread, coarse bran may be used instead.
(8) Cut into 4 in the form of a cross and bake as follows:
Plain breads – 226°C (440°F).
Fruited breads – 215°C (420°F).

Method for Farls (5 and 6) (*see* Figure 28)

(1) The Scottish Farl is usually scaled at 2 kg ($4\frac{1}{2}$ lb) for a 1,708 g (4 lb) round.
(2) Wash with a mixture of egg and milk, dust with flour or meal and dock.
(3) Divide into four in the form of a cross and bake at 215°C (420°F).

Variations

Malted Malted aerated bread can be made by adding 55 g (2 oz) of non-diastatic malt. The dough needs to be made tighter and baked at the lower temperature of 195°C (390°F).
Germ-Meal For this variety replace the wheatmeal with germ-meal and bake at 195°C (390°F).
Note If cream of tartar is used, the goods must be baked off immediately. However, if an alternative phosphate acid is used, goods can be left up to half an hour before being baked.

ERATED BREAD
00 *for details*)

Sultana Soda				Currant Soda				Scottish Farls				English Farls			
kg	*g*	*lb*	*oz*	*kg*	*g*	*lb*	*oz*	*kg*	*g*	*lb*	*oz*	*kg*	*g*	*lb*	*oz*
	500	1	2	1	000	2	4	1	000	2	0	1	000	2	4
	500	1	2		—		—		—		—		—		—
	—		—		—		—		110		4		—		—
	20		$\frac{2}{3}$		20		$\frac{2}{3}$		—		—		—		—
	30		1		30		1		—		—		—		—
	—		—						50		$1\frac{3}{4}$		50		$1\frac{3}{4}$
	85		3		85		3		—		—		55		2
	45		$1\frac{1}{2}$		45		$1\frac{1}{2}$		—		—		—		—
	—		—		—		—		—		—		30		1
	45		$1\frac{1}{2}$		45		$1\frac{1}{2}$		—		—		—		–
	15		$\frac{1}{2}$		15		$\frac{1}{2}$		15		$\frac{1}{2}$		15		$\frac{1}{2}$
	—		—		—		—		765	1	$11\frac{1}{2}$		680	1	$8\frac{1}{2}$
	915	2	1		915	2	1		—		—		—		—
	335		12		—		—		—		—		—		—
	—		—		335		12		—		—		—		—
2	490	5	$9\frac{1}{6}$	2	490	5	$9\frac{1}{6}$	1	940	4	$1\frac{3}{4}$	1	830	4	$1\frac{3}{4}$

NOVELTY BREADS

Bread can be moulded into a variety of novel shapes to provide show pieces or window attractions. The dough needs to be made very stiff so that it retains the shape into which it is moulded. Harvest Festivals provide the reason for many of these novelties, such as wheatsheafs, cornucopia (horn of plenty) loaves and fishes etc. The bread can also be coloured to provide more interest.

WHEATSHEAF

	kg	*g*	*lb*	*oz*
Medium flour	1	000	2	4
Water		445	1	0
Salt		20		$\frac{2}{3}$
Milk powder		20		$\frac{2}{3}$
Yeast		20		$\frac{2}{3}$
Totals	1	505	3	6

Fermentation time – 2 hours.
Dough temperature – 75°F.

Note For a small wheatsheaf multiply this recipe by two and for a large one, by four. This dough is timed to be ready for the oven 2 hours after it is made. Since the manipulation of the various dough pieces to form the ears of wheat, etc., is time consuming, this time must be deducted from the two hours to indicate when the modelling should commence. Even with assistance, a wheatsheaf could take an hour or more to make and during all this time the dough is proving.

Figure 29. Wheatsheaf

Method for Making Dough

(1) Make as already described for hand doughs, but ensure that all the flour is wetted before it is formed into a dough.
(2) To assist dough fermentation, it is recommended that the loose wetted flour is passed through pastry rollers until it is smooth and clear.

Method for Modelling a Wheatsheaf (*see* Figure 30).

(1) Roll out the dough to approx. 12 mm–2 cm ($\frac{1}{2}$–$\frac{3}{4}$ in) and from this cut out the base. This should be approx. half of the weight of dough used.
(2) Using a sharp knife cut the edges of the base at an angle to give a bevel. Remove sharp edges by patting with the hand.
(3) Dock the base heavily and wash with water to prevent skimming.
(4) Approx. 120 small pieces of dough are required for the ears and stalks. For a large wheatsheaf the weight for the individual ears requires to be 15 g ($\frac{1}{2}$ oz) whilst the small wheatsheaf needs ears half this size.
(5) The stalks are now rolled out and are placed onto the base so that they occupy about half the area.
(6) The ears are fashioned by first rolling small pieces of dough into a torpedo shape, and then with the scissors, making cuts along each side and down the centre.
(7) These ears are now placed onto the base in a semi-circular row starting at the top. Other ears are placed to overlap and the whole head of the wheatsheaf is thus built up until the stalks have been reached. Ensure that some of the ears are placed haphazardly to give a more natural appearance. Wash with water periodically to prevent skin formation and assist in the adhesion of the ear pieces to the base.
(8) A few strands of dough are twisted and placed from each side of the waist of the wheatsheaf and then tied in a knot in the centre.
(9) The whole wheatsheaf is now carefully washed with egg.
(10) Place into an oven at approx. 193°C (380°F). A large showpiece will take approx. $1\frac{1}{4}$–$1\frac{1}{2}$ hours to bake, whilst a small one will take 50 minutes to 1 hour.
(11) When the showpiece is baked, remove from the tray and cool on a wire mesh.

Figure 30. Method of assembling the wheatsheaf

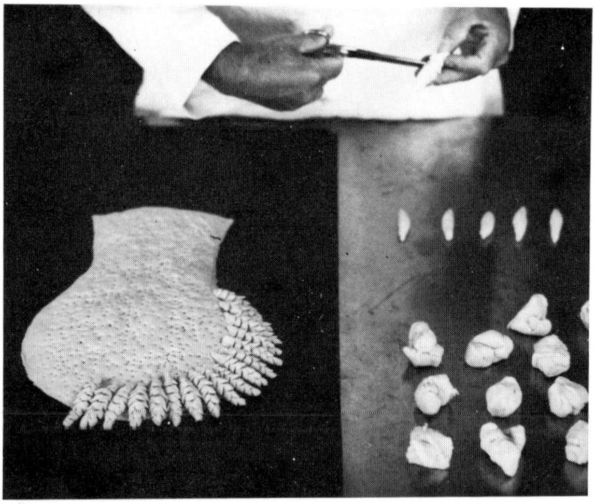

7. *Rolls – Other Varieties*

ROLLS

Figure 31. Bread rolls

These can be classified according to the meal at which they are to be consumed, as follows:

Breakfast Rolls

These are of two types, either soft or crusty. They should be unsweetened and not heavily enriched so that they may be eaten with savoury as well as sweet accompaniments.

Dinner Rolls

These should be lean, but with a crisp crust so that they may be easily broken by hand and consumed with the soup and other courses at dinner and luncheon. Vienna rolls and grissini come into this category.

ROLL
(*See pages* 105–1

	Scotch Baps				Soft				Wheatmeal			
	kg	*g*	*lb*	*oz*	*kg*	*g*	*lb*	*oz*	*kg*	*g*	*lb*	*oz*
Strong flour	1	000	2	4	1	000	2	4		—		—
Wheatmeal		—		—		—		—	1	000	2	4
Water		640	1	7		530	1	3		640	1	7
Yeast		45		1½		45		1½		45		1½
Skimmed milk powder		—		—		15		½		30		1
Salt		20		⅔		20		⅔		20		⅔
Sugar		15		½		15		½		—		—
Black treacle		—		—		—		—		15		½
Lard or fat		55		2		55		2		15		½
Butter		—		—		—		—		—		—
Egg		—		—		—		—		55		2
Bakers cheese		—		—		—		—		—		—
Dry gluten		—		—		—		—		—		—
Totals	1	775	3	15⅔	1	680	3	12⅙	1	820	4	1⅙

*Approximately

Dough temperature 24°C (76°F)
Bulk fermentation 1½ hours (Except gluten rolls)
Knock back at 1¼ hours

Tea Rolls

These are enriched, but are barely sweetened and are suitable for buffets. Bridge rolls come into this category.

Size

The size of rolls obviously depends upon personal choice, but they should be kept on the small side, particularly if served with a meal. If they are to be filled as a sandwich for a main meal, however, they can be larger.

Shape

In deciding the shape of a roll, there are two considerations:

(1) If the customer is going to slice it and spread butter, preserves or a filling before it is eaten, then the shape needs to be simple and when cut expose a good surface for spreading purposes. This would apply to most breakfast or tea rolls.

(2) If instead of it being cut, it is broken by the customer, as in a dinner roll, any shape can be made and in fact will add to the attraction of the table.

Rolls can be made from white or brown flours and can receive various treatments for variety, such as being dusted with flour, or oatmeal, or egg-washed prior to baking.

Soft Rolls

These are produced from recipes which usually contain fat and milk. The soft crust, however, is achieved by allowing them to be baked under moist conditions.

Crisp Rolls

To produce rolls with a crisp crust, the steam should be removed towards the end of the baking time and finished off in dry heat (*see* Vienna bread and rolls page 81).

details)

Bridge 1st Qual.				**Bridge 2nd Qual.**				**Cheese**				**Gluten**				**Barm Cakes**			
kg	*g*	*lb*	*oz*	*kg*	*g*	*lb*	*oz*	*kg*	*g*	*lb*	*oz*	*kg*	*g*	*lb*	*oz*	*kg*	*g*	*lb*	*oz*
1	000	2	4	1	000	2	4	1	000	2	4	1	000	2	4	1	000	2	4
	—		—		—		—		—		—		—		—		—		—
	445	1	0		445	1	0		610	1	6	1	500	3	6		625	1	6½
	50		1¾		45		1½		70		2½		—		—		45		1½
	30		1		30		1		30		1		30		1		—		—
	20		⅔		20		⅔		20		⅔		3		⅛		20		⅔
	15		½		15		½		7		¼		—		—		7		¼
	—		—		—		—		—		—		—		—		—		—
	—		—		100		3½		—		—		—		—		7		¼
	140		5		—		—		—		—		—		—		—		—
	140		5		—		—		—		—		—		—		—		—
	—		—		—		—		195		7		—		—		—		—
	—		—		—		—		—		—		890	2	0		—		—
1	840	4	12*	1	655	3	11⅙	1	932	4	5⅔	3	423	7	11⅛	1	704	3	13⅙

SPECIALITY ROLLS

Other rolls, i.e. Vienna and milk, etc., based upon bread recipes are given after the appropriate bread recipe and start on pages 79 and 81.

In the roll recipes given on pages 104 and 105 all the doughs are based upon a B.F.T. of 1 hour at 24°C (76°F) with the exception of the gluten rolls which contain no yeast.

For two-hour doughs, halve the yeast quantity given.

All the fermented doughs should receive a knock back at ¾ of their B.F.T.

Scotch Baps (Recipe on page 104)

These soft rolls are very popular in Scotland where they are often to be found on the breakfast table.

Yields 32 @ 55 g (2 oz) approx.
Yields 21 @ 85 g (3 oz) approx.

(1) Scale and mould the dough round.
(2) Allow 10 minutes to recover and pin out to an oval shape.
(3) Place onto warm baking sheets.
(4) Lightly dust with flour and bake in an oven at 250°C (480°F) for about 10 minutes (do not overbake).

Soft Rolls (Recipe on page 104)

Yields 40 @ 40 g (1½ oz) approx.
Yields 34 @ 50 g (1¾ oz) approx.
Yields 30 @ 55 g (2 oz) approx.
Yields 20 @ 80 g (3 oz) approx.

This dough may be treated in two ways:

(1) Scale at the appropriate weight, mould round, place onto a warm baking sheet, and after proving bake at 235°C (460°F).
(2) After moulding round, flatten with the rolling pin, and after proving bake in an oven at 226°C (440°F). These rolls are suitable for hamburgers.

Note For the larger rolls the baking temperature may need reducing by up to 10°C (20°F). Rolls for hamburgers need to be very soft hence the lower baking temperature.

Crusty Rolls (Recipe on page 104)

The recipe for soft rolls can be adapted for crusty rolls in the following way:
Delete fat and sugar from the recipe.
After proving, bake at the higher temperature of 247°C (480°F).
Towards the end of the baking time, withdraw oven steam and finish off in dry heat.

Wheatmeal Rolls (Recipe on page 104)

Yields 43 @ 40 g (1½ oz) approx.
Yields 37 @ 50 g (1¾ oz) approx.
Yields 32 @ 55 g (2 oz) approx.
If a dough divider is used, 36 rolls will be produced from this dough.

(1) Scale and mould either round or baton shape and place onto a warm baking sheet.
(2) Wash with egg and after proving, bake in an oven at 235°C (460°F).

Note After washing with egg, the rolls may be floured or dressed with fine oatmeal. They may also be cut before baking.

Bridge Rolls (Recipe on page 105)

	1st Quality	2nd Quality
Yields @ 20 g (¾ oz) approx.	86	78
Yields @ 25 g (1 oz) approx.	65	59

(The dough may be split into two and each piece divided in 36 by the dough divider to give a yield of 72.)

(1) Scale or divide into the appropriate number of pieces and mould round.
(2) After ten minutes recovery, mould to the shape required, i.e. finger or baton.
(3) Place onto a warm baking sheet, egg wash and place into the prover.
(4) Egg wash again at ¾ proof and bake at 232°C (450°F).

Note When egg washing, care should be exercised to ensure that no egg runs onto the tray and also that the whole roll is covered.

Cheese Rolls (Recipe on page 105)

Yields 69 @ 25 g (1 oz) approx.
Yields 46 @ 40 g (1½ oz) approx.
These may be similarly divided by the dough divider to give 72 rolls slightly under 1 oz.

(1) Incorporate the bakers' cheese at the knock-back stage.
(2) Proceed as for bridge rolls but bake in an oven at 205°C (400°F).

Note Milk rolls and Vienna rolls are dealt with on pages 79 and 81.

Gluten Rolls (Recipes on page 105)

Yield 246 @ 14 g ($\frac{1}{2}$ oz) approx.

(1) Mix the dry ingredients together and add the water.
(2) Make a thoroughly developed dough.
(3) Allow a recovery time of 30 minutes and scale into $\frac{1}{2}$-oz pieces.
(4) Mould round and place on the baking sheet.
(5) Allow another 30 minutes to relax and then bake at a temperature of 177°C (350°F).
(6) Remove steam towards the end of baking and finish off in dry heat for 10 minutes.

Note We rely solely upon the water imbibed by the extra gluten present to aerate these rolls. Therefore, it is important to make a thoroughly developed dough initially.

Barm Cakes (Recipes on page 105)

Yield 15 @ 113 g (4 oz) approx.

(1) Scale at the appropriate weight, and mould round.
(2) After ten minutes recovery, roll out and place onto a warm baking sheet.
(3) Make an indentation in the centre with the finger and prove.
(4) Place into an oven at 260°C (500°F) but turn them over half way through baking.

Note These are really muffins but are cooked in an oven instead of a hotplate. The dough is extremely soft and may require a liberal amount of flour for dusting purposes.

Aberdeen Butteries

Yield 36 @ 59 g (2 oz) approx.

	Dough				**Filling**			
	kg	*g*	*lb*	*oz*	*kg*	*g*	*lb*	*oz*
Strong flour	1	000	2	4				
Sugar		30		1				
Yeast		55		2				
Water		500	1	2				
Lard		—	—	—		225		8
Butter		—	—	—		225		8
Salt		—	—	—		20		$\frac{3}{4}$
Soft flour		—	—	—		85		3
Total dough weight	2	140	4	$12\frac{3}{4}$				

Dough @ 27°C (80°F) for 30 minutes.

(1) Make up the dough but slightly undermix.
(2) Make up the filling mixture and when the dough is ready, chop this coarsely into the dough. The aim should be to get a mass of dough in approx. 1 in cubes covered with the mixture.
(3) Dust liberally with flour and pin out to fit the bundivider.
(4) After recovery divide and place the pieces without moulding onto the baking sheet, flattening each with the hand. If no divider is available, roll out with a rolling pin, cut into strips and then squares of the appropriate weight.
(5) Prove in a cool prover for approx. 45 minutes.
(6) Bake at 246°C (475°F).

8. Bun Varieties

CHOICE OF RAW MATERIALS

Flour

All goods fermented by yeast require a strong flour, but because most recipes in this category are short-process doughs, a medium-strength flour is satisfactory for most recipes.

Fat

Low melting-point fats or oils are to be preferred so that fine dispersion throughout the dough can take place. For flavour in rich luxury bun goods, butter is to be preferred. The usual method of adding fat to the dough is by rubbing it into the flour prior to the addition of water. When made by machine, however, the fat may be softened or melted and added whilst the dough is being formed. If a large quantity of fat is in the recipe, adding it in the melted state will prevent the dough from losing its heat through the addition of a cold ingredient. A high fat content has a retarding effect upon the yeast and this might need to be increased.

Fruit

All fruit should receive a preliminary treatment before being used. This should first take the form of an inspection to see that no stones or other debris are present. The fruit should then be washed in hot water and afterwards drained of surplus moisture. During the washing and draining, the fruit will imbibe moisture which will make it fleshier.

Great care should be exercised when fruit is incorporated into a dough to ensure that no bruising takes place, since the ruptured fruit will discolour the crumb. The fruit should always be incorporated after the dough has been properly made and first warmed. Dates should be chopped finely. The use of a little flour will assist in stopping the dates sticking and effect the separation of the cut pieces.

Egg

Most rich goods contain egg which apart from its enrichment value also helps to stabilize the dough and encourages the formation of well volumed and bold goods. Because of this, the addition of egg can result in the goods being scaled at a lower temperature without any loss in the final size.

Eggs should be warmed to the temperature of the water to which they should be added for making straight doughs.

Spice

The use of this commodity, whether powdered or in liquid form, must be done with care and discretion. Powdered spice will cause discoloration of the crumb and also presents difficulty in dispersion. For this reason, liquid spice is to be preferred for doughs. Strengths vary enormously according to the quality purchased. In every case the recommendations of the manufacturer should be used or trials done to establish the right quantity to use. This commodity also has a retarding effect upon yeast which would have to be increased for spicy doughs.

Quality

Richness as far as fermented goods are concerned depends upon the amount of sugar, fat, egg and fruit that are present. If we take all the possible permutations we could construct hundreds of different recipes. In fact, when one looks at various text books we find that many of the goods for which separate recipes are given could be made equally well from merely a few selected ones. With a little technical knowledge we can take any plain dough and by adding egg, sugar, fat and fruit turn it into a rich bun dough provided we adjust the following:

Fermentation– By either increasing the B.F.T. or yeast quantity.

Consistency– By reducing the water content by the amount of egg used. Fat will also soften the dough and necessitate a reduction of the water content as a result.

The basic recipe given on page 110 represents a medium quality bun dough whilst the luxury bath bun recipe on page 118 represents about the highest quality possible.

For a leaner recipe than that given, it is only necessary to reduce the fat and sugar and delete the egg. This will affect not only the taste and crumb, but also the shape which will not be so bold. The yeast quantity could also be slightly reduced as a result.

As we have seen (page 44) the extra enrichment enjoyed by these goods has a marked retarding effect upon fermentation and the yeast quantity employed has to be adjusted to compensate for the amount of enrichment given. However, we can help make the fermentation more vigorous by introducing the yeast to these enriching ingredients through a ferment.

Ferment

This is made by mixing the optimum amount of sugar with the total liquor, yeast, milk powder and approx. $\frac{1}{10}$th of the flour quantity of the recipe. This is allowed to ferment for approx. $\frac{1}{2}$ hour before being incorporated into the other ingredients and made into a dough.

It is claimed that goods made from a ferment are superior to those made from a straight dough, but unless the enrichment is very high, most bakers would probably find the extra trouble and time involved not commercially justified by the marginal improvement to the quality of their goods.

Many advocates of the use of a ferment say that it should rise and be allowed to collapse before being incorporated into the dough. There is no justification for this practice since it depends upon many factors, i.e. shape of the vessel used, strength of flour, temperature etc. Provided the ferment has had approx. 30 minutes fermentation, this should suffice.

Basic Bun Dough

A number of different buns may be produced from a basic plain dough either by employing different finishes and/or adding other enriching agents (*see* overleaf for recipe).

Method for Ferment and Dough

(1) Take the water temperature at 76°C (170°F) less the flour temperature. This fairly high temperature is necessary to allow for the cooling of the ferment during the period at which it has to stand before being made into a dough.
(2) Whisk the milk powder into the water, disperse the yeast and whisk in the sieved flour.
(3) Cover and place in a warm place for 30 minutes.
(4) Sieve the rest of the flour and rub in the fat.
(5) After 30 minutes, whisk the rest of the ingredients into the ferment and add to the flour.

(6) Distribute the ferment throughout the dry ingredients and squeeze to form a dough. Rub this dough down on the bench for about 10 minutes until a smooth silky dough has been made which will not stick to the bench.
(7) Cover to prevent skin formation and set aside in a warm place for a further $\frac{3}{4}$ hour when the dough is ready to be made into the required varieties.

Method for a Straight Dough

(1) Sieve the flour, milk powder and salt onto the bench and into this rub the fat (or blend together with the hook on a cake machine).
(2) Place the yeast into a bowl and add about $\frac{1}{4}$ of the water with which it can be dispersed.
(3) Whisk the egg with the remaining water and add it to the flour into which you have made a bay (or on the machine).
(4) Add the yeast water.
(5) Make the dough as previously described, cover and set aside for 1 hour.

Basic Bun Recipe

Yields 38 @ 50 g ($1\frac{3}{4}$ oz) approx. – for plain varieties.
34 @ 56 g (2 oz) approx. } for fruited varieties.
27 @ 70 g ($2\frac{1}{2}$ oz) approx. }

	Ferment				Dough			
	kg	*g*	*lb*	*oz*	*kg*	*g*	*lb*	*oz*
Strong flour		100		$3\frac{1}{2}$		900	2	$\frac{1}{2}$
Water		500	1	2	—	—	—	—
Yeast		70		$2\frac{1}{4}$	—	—	—	—
Milk powder		30		1	—	—	—	—
Sugar		15		$\frac{1}{2}$		125		$4\frac{1}{2}$
Fat						100		$3\frac{1}{2}$
Egg (one)						50		$1\frac{3}{4}$
Salt						15		$\frac{1}{2}$
Total (inc. Ferment)					1	905	4	4

For a straight dough recipe, add together the amounts in both columns, i.e. Ferment and Dough.

Bun Dough Temperatures

These should be between 24–27°C (76°–80°F). If there is much processing to do to the dough to produce varieties, it is advisable to keep the temperature to the lower limit in order to reduce the risk of chilling and skin formation. The extra time involved in manipulation (e.g. Chelsea buns) will compensate for the lower temperature as far as fermentation is concerned, see page 49 for dough temperature calculations.

Knocking Back

As with bread, bun doughs also benefit from a knock back which should be given approx. $\frac{3}{4}$ of the B.F.T. (*see* page 53).

Making Bun Doughs on a Cake Machine

(1) Place the flour, fat, salt and milk powder into the machine bowl and blend in with the dough hook at bottom speed.
(2) Mix all the liquor or ferment with the sugar and egg, etc. and pour this into the flour whilst the hook is revolving.
(3) Continue at bottom speed until all the ingredients are mixed.
(4) Change to second speed and give the dough a thorough mixing for a few minutes before switching off and removing from the bowl.

Note This applies to plain unfruited doughs. Fruit is best mixed in by hand, but if done by machine it must be kept at bottom speed, otherwise some of the fruit will be ruptured and cause discolouration of the crumb (*see* page 108).

Silicone Paper

The use of silicone paper on which to bake buns is recommended where there is fruit and sugar which may make the buns stick to the tray. Also, it reduces the time necessary for cleaning the sheets between each batch being baked.

Proof

After moulding into the required shape and placed upon a baking tray, the buns should be inserted into a proving cabinet equipped with steam. For the best results the temperature of proof should be kept low – under 38°C (100°F) – and the steam kept to the minimum, sufficient to prevent a skin from forming. A hot prover with excessive steam will flatten the shape of the buns. They are usually proved to double their original size.

Glaze

For many buns it is usual to brush them as soon as they come from the oven with a wash which will form a glaze. There are three basic types used:

(1) *Proprietary* – When made and used according to the directions of the manufacturer, it will form a dry glaze on the surface of the bun.
(2) A *syrup* made as follows:

	kg	*g*	*lb*	*oz*
Sugar		500	1	2
Water		500	1	2
Gelatine (powdered)		15		½
Totals	1	015	2	4½

The water and sugar are brought to the boil, any scum removed, and the gelatine added dispersed in a little water. This gives a sweet but sticky glaze ideally suited to Chelsea buns.

(3) An *egg custard* made by beating two or three eggs into ½ litre or 1 pint of milk (the quantities are not critical). This glaze is also dry.

For additional flavour salt may be added to these glazes.

Cooling

Buns should be removed from the baking sheet immediately they are baked and placed upon cooling wires. If left on the sheet to cool, steam will condense on the underneath to give a soggy base to the bun. If left for a prolonged period rust will form which will come off and stain the underside of the bun.

VARIETIES FROM BASIC BUN DOUGH

Figure 32. Varieties of fermented buns. From *left* to *right* (1) Chelsea (2) Bath (3) Currant (4) Doughnuts (5) Swiss (6) Cream

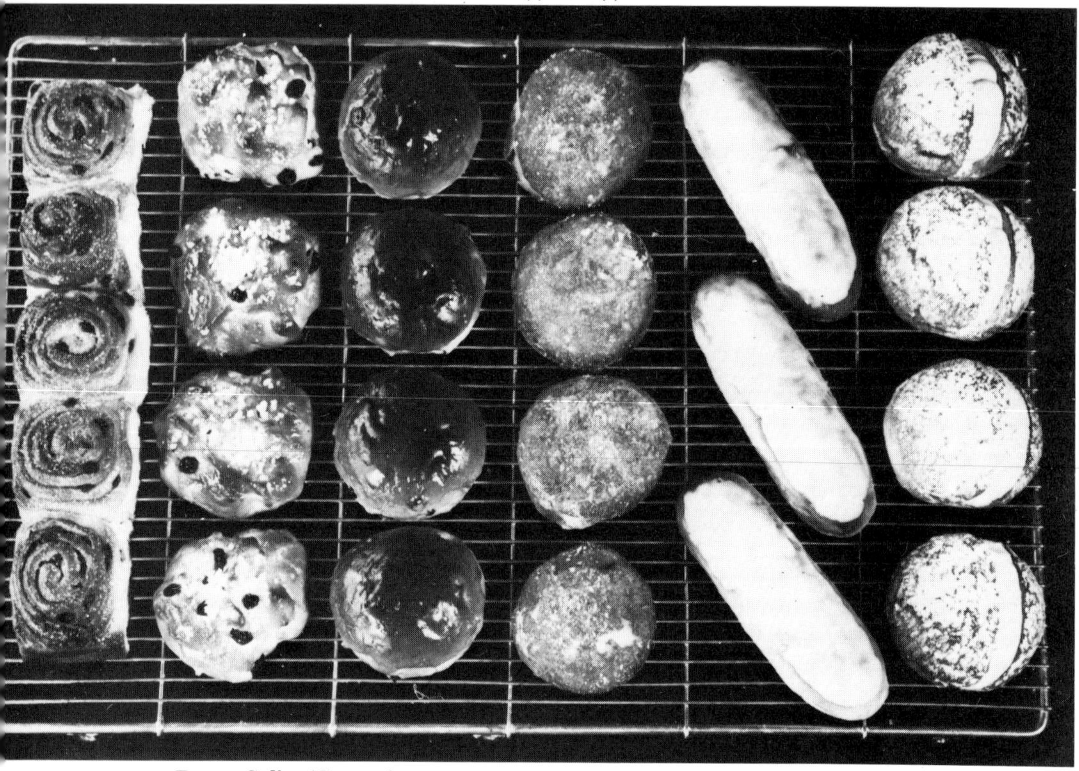

Devon Splits (Cream buns or cream cookies) (*see* Figure 32 (6))

Allow 6 decilitres (1 pint) fresh cream.

(1) Scale the dough and mould each piece into a ball.
(2) Place onto a clean baking sheet.
(3) Prove until approx. twice the initial size.
(4) Bake in an oven at 232°C (450°F) for 12–15 minutes.
(5) When cold cut a slit $\frac{3}{4}$ through the top at an angle.
(6) Pipe in a bulb of raspberry jam.
(7) Follow with a bulb of whipped fresh cream or filling cream.
(8) Liberally dust with icing sugar.

Butter Buns

(1) Scale and mould round as for Devon splits.
(2) Roll out with a rolling pin to approx. 3 mm ($\frac{1}{8}$ in) in thickness.
(3) Brush melted butter over the surface and fold in half.
(4) Repeat (3) so that we now have quarter circles. For variety, jam, curd or softened macaroon paste may be piped into the centre of each bun before folding.
(5) Place the buns to form a round on the tray.
(6) Brush over with egg wash or milk and fully prove.
(7) Put into an oven at 238°C (460°F) and when baked dredge lightly with castor sugar.

Swiss Buns (Iced Fingers) (*see* Figure 32 (5))

(1) Scale and mould the pieces into a ball.
(2) After ten minutes recovery, elongate the pieces to form fingers.
(3) Place these on a clean baking tray and after proof bake at 232°C (450°F) for approx. 12 minutes.
(4) When cold, dip the top into water icing which may be flavoured and coloured as desired (*see* recipe on page 141).

Spiced Fingers

Yield – 48 slices.

	kg	*g*	*lb*	*oz*
Basic bun dough	1	905	4	4
Butter		55		2
Spice		7		$\frac{1}{4}$
Sultanas		450	1	0
Totals	2	417	5	$6\frac{1}{4}$

(1) Roll out the ripe basic bun dough to cover an area of 42 cm × 60 cm (15 in × 24 in).
(2) Melt the butter and brush this over $\frac{2}{3}$rds of the dough (lengthwise) with the exception of the bottom edge which is egg washed to seal the piece after folding.
(3) Sprinkle the sultanas over the melted butter and very lightly sprinkle on the mixed spice. To help achieve an even spread, this can be first mixed with a little sugar.
(4) Fold the untreated dough surface into the centre and fold again puff pastry fashion to give 3 layers of dough and two of filling.
(5) Using the rolling pin, extend the dough to cover an area of 40 cm × 45 cm (16 in × 18 in).
(6) Egg wash and then cut into 48 fingers, approx. 10 cm × 4 cm (4 in × $1\frac{1}{2}$ in).
(7) Place onto a warm baking sheet and after proving, bake in an oven at 216°C (420°F).

Currant Buns (*see* Figure 32 (3))

Basic bun dough recipe.
Currants – 310 g (11 oz).
Yields 40 @ 55 g ($2\frac{1}{2}$ oz).
Yields 32 @ 70 g (2 oz).

(1) Carefully mix the currants into the bun dough recipe so that no bruising occurs, and then proceed as for Devon Splits.
(2) When baked, brush over with a glaze immediately they are withdrawn from the oven.

Small Teacakes (Figure 33) (*For Yorkshire Teacakes see page* 120).

These may be made from either the plain or the fruited dough.

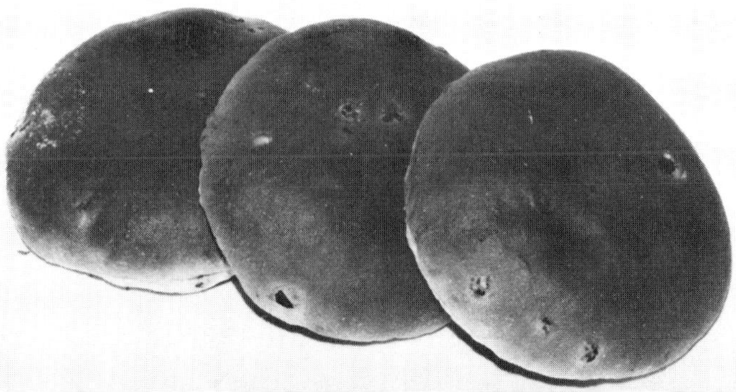

Figure 33. Tea cakes

(1) Scale and mould round as for currant buns.
(2) Allow 10–15 minutes recovery and then roll with the rolling pin to about 6 mm ($\frac{1}{4}$ in) thick. They may either be round or oval in shape.
(3) Set out on a greased or silicone tray and dock each one.

At this stage there are a number of different treatments which may be given to produce alternative varieties.

Variety A

(1) Egg wash, fully prove and bake in an oven at 226°C (440°F).
(2) Once set in the oven and the teacakes have started to colour, withdraw, and with a large palette knife turn them over.
(3) Return to the oven to finish baking.
(4) On removal reverse them again.

Variety B

(1) Do not egg wash, but leave plain.
(2) Turn in the oven as for the previous variety.

Note If desired, a light dusting of flour may be given to the pieces before baking.

Variety C

(1) Leave plain and do not turn them over in the oven.

Bun Rounds

Use either the plain or fruited dough.

Variety A

(1) Proceed as for Devon Splits.
(2) Place the moulded pieces as a ring onto a baking sheet or in a cottage pan. It is usual to form a ring of seven with one in the centre, i.e. eight if placed on a tray. When baked in a cottage pan a round of six pieces may be made.
(3) Egg wash, prove and bake in an oven at 232°C (450°F).

Variety B (*see* Figure 34 (2) below)

(1) Scale the dough according to the size required. For a round divided into four 170 g (6 oz) dough is usually taken.
(2) Mould into a ball and allow 10 minutes recovery time.
(3) Roll out with a rolling pin to approx. 12 mm ($\frac{1}{2}$ in) in thickness and place upon the baking tray.
(4) Using a Scotch scraper make cuts to divide the round into 4, 6 or 8 segments according to the size required.
(5) Egg wash, prove and bake in an oven at 232°C (450°F).

Variety C (*see* Figure 34 (1))

(1) Proceed as for Devon Splits.
(2) Mould the pieces into a pear shape and assemble them on the tray in the form of a round of six, seven, or eight with the pointed end towards the centre.
(3) Egg wash, prove and bake at 232°C (450°F).

Note Varieties B and C may have a spot of jam placed in the centre and also be dusted with castor sugar prior to baking.

Florentine Buns

Yield—13 buns.

	kg	*g*	*lb*	*oz*
Basic bun dough	1	905	4	4
Florentine mixing*		585	1	5
Whipped fresh cream or custard		140		5
Totals	2	630	5	14

(1) Scale the dough at 140 g (5 oz) and mould round.
(2) After 10 minutes recovery, roll the pieces to a flat disc of approx. 18 cm (7 in) diameter and place into slightly greased cottage pans of this size.
(3) Spread 45 g ($1\frac{1}{2}$ oz) of florentine mixing over the top to within 2 cm ($\frac{3}{4}$ in) from the edge.
(4) After proving, bake at 204°C (400°F).
(5) When cool, cut in half horizontally and cover the lower half with whipped cream or custard.
(6) Cut the top piece into eight and replace each segment onto the creamed base.

Chelsea Buns (*see* Figure 32 (1))

Yields 43 @ 55 g (2 oz) approx.
or 34 @ 70 g ($2\frac{1}{2}$ oz) approx.

	kg	*g*	*lb*	*oz*
Basic bun dough recipe	1	905	4	4
Currants		400		$14\frac{1}{2}$
Brown sugar		55		2
Mixed spice		7		$\frac{1}{4}$
Cooking oil		55		2
Totals	2	422	5	$6\frac{3}{4}$

(1) Roll out the basic dough to cover an area of approx. 38 cm (15 in) by 60 cm (24 in).
(2) Mix the brown sugar and spice together, spread this mixture over the dough and sprinkle over with the currants.
(3) Roll up swiss roll fashion.
(4) Cover the surface with oil.
(5) Cut into the appropriate number of slices.
(6) Place each slice flat side down onto a well greased tray with the pieces almost touching.

See* companion volume, **Bakery: Flour Confectionery

(7) After proving bake in an oven at 226°C (440°F) for approx. 12–15 minutes.
(8) When baked and whilst still hot, brush over liberally with the syrup glaze and dust with castor sugar.

Notes:

(1) Traditionally these goods should be so soaked with syrup that they will squelch when compressed with the fingers.
(2) Instead of placing these goods on a baking tray they can be placed into foil dishes to be proved, baked and sold. Usually these are designed for units of six in which case a choice has to be made as to whether the size is increased to give a yield of seven trays 6 × 7 = 42, or six trays 6 × 6 = 36.

Bath Buns (*see* Figure 32 (2))

Yields 42 @ 55 g (2 oz) approx.
or 33 @ 70 g ($2\frac{1}{2}$ oz) approx.

	kg	*g*	*lb*	*oz*	
Basic bun recipe	1	905	4	4	
Sultanas		300		$10\frac{1}{2}$	
Peel		50		$1\frac{3}{4}$	
Sugar nibs		100		$3\frac{1}{2}$	for the dough
Sugar nibs		30		1	for top decoration
Totals	2	385	5	$4\frac{3}{4}$	

(1) Mix the fruit and sugar nibs into the basic dough, taking care not to bruise the sultanas.
(2) Divide into the appropriate number of pieces and place them upon a baking sheet in a rough shape.
(3) Prove and before placing them into the oven sprinkle a few sugar nibs upon each.
(4) Bake in an oven at 226°C (440°F) for approx. 12–15 minutes.
(5) Immediately they are removed from the oven, glaze as for currant buns.

Doughnuts (*see* Figure 32 (4))

Figure 35. Jam injector. This is operated by pressing down a lever to inject jam through a nozzle at the top on which is impaled the doughnut. The quantity of jam injected can be altered to suit individual requirements.

(1) Proceed as for Devon Splits.
(2) When moulded, insert the thumb to leave an impression which is filled with raspberry jam and sealed afterwards. Alternatively, this jam may be inserted by a jam injector after the doughnuts are cooked. They may also be split and filled with cream after baking, in which case they are left plain.
(3) Place on an oiled tray to prove.
(4) When fully proved, drop into hot fat at a temperature of approx. 193°C (380°F).
(5) The pieces will float and it will be necessary to turn them over to ensure both sides are cooked to a golden brown colour. Special doughnut fryers are available with a wire grid which can be placed over the dough pieces to depress them below the level of the hot fat. This eliminates the need to turn them over and a more uniform colour and cooking results.
(6) Remove, drain off excess fat and then roll in castor sugar to which a little cinnamon spice has been added.

Notes:
(1) The type of fat or oil used for doughnut frying is important. It should have a low smoke-point and low absorption characteristics. The advice of a reputable fat manufacturer should be accepted as to the choice available.
(2) The correct frying temperature is also very important. Too low a temperature will result in greater fat absorption and greasy doughnuts, too high a temperature will result in either having to remove the doughnuts before they are cooked or leaving them to get too coloured, also the useful life of the fat is shortened.

Other shapes

Rings Roll out the dough to approx. 12 mm ($\frac{1}{2}$ in) in thickness and cut out rings using two cutters of different dimensions.
Fingers After moulding round extend to form a finger shape.

Other fillings
Lemon – Lemon Curd.
Almond – Almond fillings on pages 138 and 139.
Custard – Custard filling on page 140.
Cream – Either dairy or a suitable filling cream.
Fruit – Pineapple and ginger crush. Tinned fruits (may be used with cream).

Other finishes
(1) Dip tops in a suitably flavoured water icing or fondant, e.g. lemon.
(2) Dip the whole doughnut into a thinned icing and allow to drain.
(3) Dip into a boiling sugar solution at 115°C (240°F) and then roll in roasted nib or flaked almonds or desiccated coconut.
(4) Dip the whole doughnut into the following glaze and allow to drain.

	kg	*g*	*lb*	*oz*
Hot water 38°C (100°F)	1	000	2	4
Icing sugar	3	560	8	0
Fondant		335		12
Agar agar		4		$\frac{1}{8}$
Concentrated fruit juice		210		$7\frac{1}{2}$
Totals	5	109	11	$7\frac{5}{8}$

Luxury Bath Buns

Yields 54 @ 55 g. (2 oz) approx.
Yields 43 @ 70 g. (2½ oz) approx.

	Ferment				Dough				
	kg	*g*	*lb*	*oz*	*kg*	*g*	*lb*	*oz*	
Strong flour		55		2					A
Water		290		10½					
Egg		290		10½					
Yeast		55		2					
Sugar		30		1					
Strong flour						945	2	2	B
Salt						7		¼	
Butter						450	1	00	
Sultanas						285		10	
Mixed peel						55		2	
Nib sugar						450	1	00	C
Nib sugar						110		4	D
Weight of total dough including ferment					3	022	6	12¼	

Dough – 1 hour at 24°C (76°F) approx.

(1) Mix the egg and water together and raise its temperature to approx. 38°C (100°F) before incorporating the other ingredients to form the ferment (A) set aside well covered for approx. ½ hour.
(2) Add the ferment to the warmed ingredients at B in two stages:
   (*a*) Firstly make a well developed dough by excluding the fruit.
   (*b*) Carefully blend in the fruit without bruising.
(3) After 1 hour's B.F.T. add the nib sugar at (C) and mix well into the dough.
(4) Allow a recovery period of 15 minutes, during which time a baking sheet should be greased with butter.
(5) Divide and lay the pieces on the buttered tray, keeping them as rocky as possible.
(6) Wash with egg (optional).
(7) Prove *in the cold* but in a moist atmosphere.
(8) Sprinkle some of the nib sugar at (D) onto each piece.
(9) Bake in an oven at a temperature of 221°C (430°F) for approx. 15 minutes.
(10) Lightly brush the pieces with bun wash immediately they are removed from the oven.

Hot Cross Buns (*see* Figure 36)

These traditional buns are usually made from a good quality spiced recipe and are sold at Eastertide. Each bun is decorated with a cross to mark the religious festival at this time.

There are four ways by which this cross may be marked upon the bun:

(*a*) Halfway through the final proof a wooden block in the shape of a cross is pressed into each bun to leave an impression. This method is not recommended because the bun loses its bold shape and the impression of the cross is not very distinctive.

Figure 36. Hot cross buns with piped crosses

(*b*) Metal crosses are laid onto each bun half way through the final proof. The metal prevents that portion underneath from taking an oven colour, so leaving a white cross.
(*c*) Piping a thin paste using a No. 3 piping tube (approx. 3 mm ($\frac{1}{8}$ in)) in the form of a cross over the bun after proof.
(*d*) Placing wafer crosses on top of the bun at $\frac{1}{2}$ proof stage.
The latter method has been patented and is used extensively in the baking industry. Special dispensers are available to deposit the wafer cross onto each bun.

Yield 64 @ 40 g ($1\frac{1}{2}$ oz) approx., or 48 @ 55 g (2 oz) approx.

	kg	*g*	*lb*	*oz*
Strong flour	1	000	2	4
Water		555	1	4
Milk powder		55		2
Yeast		85		3
Butter or fat		195		7
Sugar		195		7
Egg		220		8
Salt		7		$\frac{1}{4}$
Currants		220		8
Sultanas		110		4
Mixed peel		55		2
Totals	2	697	6	$1\frac{1}{4}$

B.F.T.–1 hour @ 24°C (76°F).

Spice

This may be added either in liquid form or as ground spice. The former needs careful dispensation to ensure that it is not used in excess. Ground spice will colour the crumb but is easier to measure in small quantities. If this is used, brown sugar may be employed in the recipe for added flavour.

Bun Cross Paste

	kg	*g*	*lb*	*oz*
Soft flour	1	000	2	4
Water	1	110	2	8
Shortening		220		8
Baking powder		7		¼
Milk powder		110		4
Salt		7		¼
Totals	2	454	5	8½

(1) Scale the dough and mould the pieces as for currant buns.
(2) Place the moulded buns upon the baking sheet in rows to facilitate the application of the crosses and place in the prover.
(3) After half proof (or full proof for method (C)) apply the cross as previously described.
(4) When fully proved, bake in an oven at 246°C (475°F).
(5) Immediately they are baked, wash with a suitable bun wash.

Notes Since Hot Cross Buns are traditionally made once a year at Eastertide, there is usually great pressure to produce enough buns in excess of the normal production schedule, to satisfy demand. The facility of deep freezing offers a solution to this problem, in that buns may be made well in advance, stored in deep freezers and defrosted as and when required.

The use of a good quality enriched bun dough is essential for this process.

Yorkshire Teacakes (*see* Figure 33)

These are traditional goods which may be made from either white, wholemeal or wheatmeal flour. There are two types:

Plain These are not very sweet and are usually split, buttered and used with savoury and meat fillings.

Fruited These are sweet and with plenty of fruit. They are often split and toasted before being buttered.

Plain Teacake recipe

Yield 17 @ 100 g (3½ oz) approx.
or 15 @ 115 g (4 oz) approx.

	kg	*g*	*lb*	*oz*
Strong flour	1	000	2	4
Water		550	1	4
Yeast		55		2
Milk powder		30		1
Sugar		15		½
Fat		70		2½
Salt		15		½
Totals	1	735	3	14½

B.F.T. – 1 hour at 25°C (76°F).

For Fruited Varieties

Add – Yeast 15 g ($\frac{1}{2}$ oz)
Sugar 85 g (3 oz)
Fruit 225 g (8 oz)

Reduce – Salt by 7 g ($\frac{1}{4}$ oz).

For Wholemeal and Wheatmeal

Replace – White flour for wholemeal or wheatmeal

Add – 30 g (1 oz) extra water.

(1) Scale the dough at the required size and mould round.
(2) Roll out the pieces to a diameter of approx. 10 cm (4 in) with the rolling pin.
(3) Place upon the baking sheet and dock each piece in the centre.
(4) Brush with diluted egg wash or milk.
(5) Prove and bake at the following temperatures:
Plain – 249°C (480°F)
Fruited – 238°C (460°F).

Notes It is important that the prover should not be too hot or too full of steam as this will result in the goods having surface blisters with holes forming under the top crust.

9. Teabreads

REGIONAL VARIETIES OF TEABREAD

HUFFLERS

B.F.T. $2\frac{1}{2}$ hours @ 21°C (70°F).
Knock back at 2 hours.

	kg	*g*	*lb*	*oz*
Strong flour	1	000	2	4
Milk powder		30		1
Lard or shortening		35		$1\frac{1}{4}$
Sugar		10		$\frac{1}{3}$
Salt		20		$\frac{2}{3}$
Yeast		30		1
Water		680	1	$4\frac{1}{2}$
Totals	1	805	3	$12\frac{3}{4}$

Essex Hufflers (Yield 2 rounds – 16 segments)

(1) Scale the ripe dough into 900 g (1 lb 14 oz) pieces.
(2) Mould round and after recovery roll out with a rolling pin.
(3) Place upon the baking sheet, egg wash and dust liberally with flour.
(4) Cut into eight segments as in a bun round.
(5) After proving, they are baked in an oven at 230°C (450°F) and then broken into eight segments.

Kentish Huffkins (Yield 30 at 60 g (2 oz); 20 at 90 g (3 oz))

(1) Scale the ripe dough at 20 g (2 oz) or 30 g (3 oz) and mould round.
(2) After recovery, pin out slightly.
(3) Place onto the baking tray and dust with flour.
(4) Prove and bake in an oven at 230°C (450°F).

BUN LOAVES

These are made from an ordinary bun dough scaled at 340 g–370 g (12–13 oz) and placed into small bread tins to prove and bake.

The currant bun recipe on page 110 may be used with slight adjustments and the straight dough for this is given here.

Yield 6 loaves scaled @ 360 g (13 oz) approx.

	kg	*g*	*lb*	*oz*
Strong flour	1	000	2	4
Water		500	1	2
Yeast		60		$2\frac{1}{4}$
Milk powder		30		1
Eggs		30		1
Sugar		85		3
Fat		85		3
Salt		15		$\frac{1}{2}$
*Fruit *see below*		400		$14\frac{1}{2}$
Totals	2	205	4	$15\frac{1}{4}$

B.F.T. 1 hour @ 24°C (76°F).

*Fruit

This may be either sultanas or currants or a mixture of both. Up to 30 g (2 oz) of peel may be used.

(1) Make the dough in the usual way but omit the fruit.
(2) Knock back at $\frac{3}{4}$ B.F.T. and incorporate the *warm* fruit.
(3) Scale at the appropriate weight in accordance with the tin used.
(4) Mould first round and then elongate to fit the tin.
(5) Prove in minimum steam and at low heat.
(6) Egg wash and bake in an oven at 204–215°C (400–420°F) according to size for approx. 35 minutes.
(7) Wash with a glaze immediately the buns are withdrawn from the oven.
(8) Remove from the tin and set on cooling wires to prevent sweating.

Sally Lunns (Yield 8 from basic bun dough recipe)

These traditional goods are made from a good quality bun dough, but baked in a hoop as follows:

(1) Make a dough from the basic bun dough recipe. This can be further enriched by adding another egg 50 g ($1\frac{3}{4}$ oz) if desired.
(2) Knock back at $\frac{3}{4}$ B.F.T. and scale at approx. 225 g (8 oz).
(3) Mould round and leave to recover.
(4) Remould into a round shape, flatten slightly and egg wash.
(5) Drop the pieces into a greased and warmed cake hoop size 13 cm (5 in) on a baking sheet.
(6) Give a cool proof in a moist atmosphere and bake at 227°C (440°F).
(7) Remove from the hoops immediately they are baked.

Several finishes may be made as follows:

(*a*) Dusted with icing sugar.
(*b*) Iced with fondant.
(*c*) Bun washed and dipped into castor sugar.

Note These goods are often eaten in toasted slices with butter and preserves.

TRADITIONAL BUN LOAVES

A review of the current literature available reveals a large number of traditional recipes which are very enriched bun doughs, but on analysis can all be made from a basic dough.

The problem with all enriched doughs is that the extra enriching agents such as sugar and fat and the spices which some contain, has such a retarding effect upon the yeast that ripeness and gassing is seriously delayed. The solution to this is to make a plain dough which can readily ferment and then add the enriching ingredients after it has become ripened with fermentation.

Proof A cool moist proof is required. This might be prolonged up to 1½ hours according to the variety.

Baking The oven temperature will depend upon the amount and type of enriching agents present in the dough, but will generally be cooler than for most bun doughs. Time of baking will depend upon the size, the larger loaves taking up to 1½ hours to bake in a cool oven.

Cooling After baking, remove immediately and place onto wires to cool.

All the following varieties can be made this way:

Name	*Origin*
Bara Brith	Wales
Barm Brack	Ireland
Lincolnshire Plum Loaf	Lincolnshire
Selkirk Bannocks	Scottish border
Guernsey Gauche	Channel Islands
Dough Cake	West Country
Saffron Loaf	Cornwall

TRADITIONAL BUN LOAV

(See ab

	Bara Brith				**Barm brack**				**Lincolnshire Plum Loaf**			
	kg	*g*	*lb*	*oz*	*kg*	*g*	*lb*	*oz*	*kg*	*g*	*lb*	*oz*
Butter		220		8		—		—		—		—
Lard		—		—		150		5½		220		8
Egg		140		5		140		5		—		—
Mixed spice		15		½		—		—		—		—
Nutmeg		—		—		—		—		2		$\frac{1}{16}$
Ginger		—		—		—		—		2		$\frac{1}{16}$
Sugar		180		6½		100		3½		220		8
Currants		780	1	12	1	225	2	12		970	2	6½
Sultanas		280		10		—		—		220		8
Raisins		280		10		—		—		140		5
Cut mixed peel		140		5		165		6		60		2½
Orange peel		—		—		—		—		—		—
Saffron		—		—		—		—		—		—
Water		—		—		—		—		—		—
Wt. of filling	2	035	4	9	1	780	4	0	1	834	4	6⅛
Total weight with dough	3	695	8	4½	3	440	7	11½	3	494	8	1⅝

BASIC DOUGH

	kg	g	lb	oz
Strong flour	1	000	2	4
Water		555	1	4
Yeast		60		2
Salt		15		½
Milk powder		30		1
Totals	1	660	3	11½

B.F.T. 1 hour @ 24°C (76°F).

To this basic dough the ingredients listed below are added in the following way:

(1) Mix the ingredients under A and making sure that they are warm, incorporate them into the ripened dough. Thoroughly mix to form a clear dough with no lumps or stickiness.

(2) Warm the fruit of B and carefully incorporate making sure it is well mixed into the dough without any bruising.

Each variety is now treated in a different manner as follows:

Bara Brith (Yield 8 @ 445 g (1 lb); 7 @ 510 g (1 lb 2 oz)

(1) Scale dough at the appropriate weight, mould and place in a suitable sized bread tin.

(2) Brush over with a diluted egg wash and prove.

(3) Bake at a temperature of 193°–204°C (380°–400°F).

ROM A BASIC DOUGH

r details)

Selkirk Bannocks				Guernsey Gauche				Dough Cake				Saffron Loaf				
kg	g	lb	oz	kg	g	lb	oz	kg	g	lb	oz	kg	g	lb	oz	
	150		5½		390		14		500	1	2		250		9	
	150		5½		195		7		—		—		—		—	
	—		—		—		—		55		2		—		—	
	—		—		—		—		15		½		—		—	A
	—		—		3		1/10		—		—		—		—	
	—		—		—		—		—		—		—		—	
	250		9		—		—		125		4½		125		4½	
	—		—		765	1	11½		445	1	0	1	000	2	4	
	890	2	0		—		—		445	1	0		—		—	
	125		4½		—		—		—		—		—		—	B
	—		—		195		7		70		2½		110		4	
	—		—		125		4½		—		—		—		—	
	—		—		—		—		—		—		7		¼	
	—		—		—		—		—		—		30		1	
1	565	3	8½	1	673	3	12 1/10	1	655	3	11½	1	522	3	6¾	
3	225	7	4	3	333	7	7⅗	3	315	7	7	3	182	7	2¼	

Barm Brack (Yield 5 @ 680 g ($1\frac{1}{2}$ lb))

(1) Scale dough at the appropriate weight, mould round and slightly flatten.
(2) Wash with egg and place into warm greased 14 cm ($5\frac{1}{2}$ in) diameter cake hoops on a baking sheet.
(3) Prove and bake in an oven at 215°C (420°F).

Lincolnshire Plum Loaf (Yield 8 @ 445 g (1 lb))

(1) Scale dough at the appropriate weight, mould and place into suitable size bread tins.
(2) Wash with egg and prove.
(3) Bake in an oven at 205°C (400°F).

Selkirk Bannocks (Yield 7 @ 445 g (1 lb))

(1) Scale the dough at the appropriate weight and mould round.
(2) After ten minutes recovery, remould, slightly flatten and place upon warmed baking sheets.
(3) Egg wash and after proof, baking in an oven at 215°C (420°F).

Guernsey Gauche

This bun loaf is made into large loaves and is cut up and sold in the shops by weight. It is usually baked in tins having sloping sides similar to domestic oven tins.

(1) Shape and place the dough into the well greased tin.
(2) After $\frac{1}{2}$ hour proof, fold the dough back in the tin and turn it over to leave a smooth skin on top.
(3) Prove for a further 15–20 minutes and bake in an oven at 195°C (380°F) for approx. $1\frac{1}{2}$ hours.

Other Varieties

Sultana Gauche

Replace – Currants with 1 kg ($2\frac{1}{4}$ lb) sultanas.

Butter Gauche

Replace – Lard with Butter.
Replace – Mixed Peel with 65 g ($2\frac{1}{2}$ oz) well chopped lemon peel.
Add – 65 g ($2\frac{1}{2}$ oz) sugar.

Dough Cake (Yield 8 @ 410 g ($14\frac{1}{4}$ oz))

Any ripened bread dough may be used instead of the basic recipe given.

(1) Scale at the appropriate weight for the size of tin used. These can be of any shape. Alternatively, hoops may be used as for Sally Lunns.
(2) Prove and bake at a temperature of 193°–204°C (380°–400°F).
(3) After baking and whilst still hot, brush the surface of the loaves with melted lard or fat.

Saffron Loaf

Although not strictly a bun loaf, the saffron loaf is included here because it can be made from the same basic recipe. The saffron has to be first made into an infusion with the 30 g (1 oz) of boiling water before being added to the dough. Besides colouring the dough yellow, the saffron also adds flavour which is favoured by the natives of Cornwall.

LARDY CAKES

This is made by folding a mixture of lard and sugar into a ripened bun dough. The three varieties given are all made from the following basic dough:

Basic Dough

	kg	*g*	*lb*	*oz*
Strong flour	1	000	2	4
Water		585	1	5
Yeast		60		2
Milk powder		30		1
Sugar		70		2½
Lard		55		2
Salt		15		½
Totals	1	815	4	1

B.F.T. 1 hour @ 24°C (76°F).

Oxford Lardy

(Yield 6 @ 400 g (14 oz)

	kg	*g*	*lb*	*oz*
Basic dough	1	815	4	1
Lard		390		14
Brown sugar		195		7
Mixed spice		7		¼
Totals	2	407	5	6¼

(1) Mix the lard, sugar and spice to a cream.
(2) Roll out the dough about 30 cm × 60 cm (1 ft × 2 ft) and spread the above mixture over ⅔rds of the surface. Fold the plain section into the centre and the remaining creamed section over to form two layers of filling and three of dough.
(3) Proceed to give the dough three ½ turns (*see* Danish pastry page 137).
(4) Allow the dough to recover and roll out to a size sufficient so that six pieces can be cut which will fit into shallow farmhouse tins.
(5) Grease the tins well, cut the dough into the six pieces and lay flat into the tins.
(6) Brush over with diluted egg wash and with a knife cut a trellis pattern on top.
(7) After proof bake in an oven at 205°C (400°F).

Wiltshire Lardy

Yield one full size baking sheet which may be cut into suitable sized pieces, i.e. 15 cm × 15 cm (6 × 6 in) = 18 pieces. $11\frac{1}{2}$ cm × $11\frac{1}{2}$ cm ($4\frac{1}{2} \times 4\frac{1}{2}$ in) = 32 pieces.

	kg	*g*	*lb*	*oz*
Basic dough	1	815	4	1
Currants		390		14
Lard		780	1	12
Brown sugar		290		$10\frac{1}{2}$
Mixed spice		7		$\frac{1}{4}$
Totals	3	282	7	$5\frac{3}{4}$

(1) Add the currants to the dough at the knock back stage ($\frac{3}{4}$ of B.F.T.).
(2) Proceed as for Oxford Lardy – methods 1, 2 and 3.
(3) Allow the dough to recover and then roll out to cover a full sized well greased 30 × 18 in baking sheet (75 cm × 46 cm approx.).
(4) Brush over with diluted egg wash and mark a trellis pattern on with a knife.
(5) Prove and bake at 205°C (400°F).
(6) On removal from the oven, cut into the appropriate sized pieces.

Gloucester Lardy

Yield 6 @ 455 g (1 lb).

	kg	*g*	*lb*	*oz*
Basic dough	1	815	4	1
Currants		390		14
Sultanas		140		5
Lard		195		7
Brown sugar		195		7
Totals	2	735	6	2

(1) Add the fruit to the basic dough at the knock back stage ($\frac{3}{4}$ B.F.T.).
(2) Proceed as for Oxford Lardy – methods 1 and 2.
(3) Roll out the dough to its original size and roll up swiss roll fashion from the extended end. This should give a roll approx. 30 cm (12 in) in length.
(4) Cut into six, giving slices of approx. 5 cm (2 in) and place with the cut side down into well greased cottage pans.
(5) After proving bake in an oven at 205°C (400°F).
(6) When baked, remove carefully and display upside down to show the caramelized surface which has been formed.

FERMENTED CAKE

These are similar to the bun loaves and the final product looks very similar. The main difference is in the crumb which tends to eat more like cake. This is due to it being partially aerated with baking powder.

Two recipes are given here both made from a basic ferment.

Ferment

(1) Whisk the milk powder into the water or alternatively omit and use fresh milk.

FERMENTED CAKE

(See page 128 for details)

	Ferment				Old English				Liverpool Currant				
	kg	*g*	*lb*	*oz*	*kg*	*g*	*lb*	*oz*	*kg*	*g*	*lb*	*oz*	
Strong flour		110		4									
Water		280		10									
Egg		280		10									
Milk powder		30		1									
Sugar (castor)		30		1									
Yeast		45		$1\frac{1}{2}$									
Strong flour						890	2	0		890	2	0	
Salt						7		$\frac{1}{4}$		15		$\frac{1}{2}$	
Baking powder						30		1		30		1	
Mixed spice						—		—		15		$\frac{1}{2}$	A
Ground ginger						7		$\frac{1}{4}$		—		—	
Ground cinnamon						7		$\frac{1}{4}$		—		—	
Ground nutmeg						7		$\frac{1}{4}$		—		—	
Sugar (castor)						445	1	0		—		—	
Brown sugar						—		—		530	1	3	
Butter or margarine						445	1	0		—		—	
Lard or fat						—		—		375		$13\frac{1}{2}$	B
Egg						—		—		140		5	
Treacle						—		—		250		9	
Caramel colour						—		—		as		req.	
Currants						305		11		750	1	11	
Sultanas						150		$5\frac{1}{2}$		555	1	4	C
Mixed peel						—		—		150		$5\frac{1}{2}$	
Lemon peel						55		2		—		—	
Total weight of doughs					3	123	7	0	4	475	10	$0\frac{1}{2}$	

(2) Add the egg and bring the mixture to a temperature of approx. 38°C (100°F).
(3) Whisk in the yeast and lastly the flour.
(4) Cover and set aside for half an hour to ferment.

Dough

(1) Sieve the dry ingredients at A.
(2) Add to the ferment and mix to a dough.
(3) Make a cream of the ingredients at B and gradually incorporate these into the dough.
(4) Once a clear dough has been made, add the ingredients at C and mix in well.

Old English

(Yield 7 @ 445 g (1 lb) in a 13 cm (5 in) hoop, or 5 @ 624 g (1 lb $6\frac{1}{2}$ oz) in a 14 cm ($5\frac{1}{2}$ in) hoop)

(1) Scale into papered square pans or hoops, the size depending upon the weight chosen.
(2) After proving a little in a cool prover, bake in an oven at 193–204°C (380–400°F).

Liverpool Currant (Yield 5 @ 890 g (2 lb))

(1) Scale into well greased large bread tins.
(2) After a cool proof, bake in an oven at 193–204°C (380–400°F) for approx. 1 hour.

Scotch Black Bun

This traditional bun from Scotland is a very richly fruited article which is usually offered to guests as an accompaniment to the national drink – Whisky. Made properly it may be kept up to a year if correctly stored and because it contains spirit as well as fruit, the flavour is improved on keeping. It consists of a thin dough case in which the very rich filling is baked.

	BASIC		DOUGH	
	kg	g	lb	oz
Strong flour	1	000	2	4
Salt		20		⅔
Yeast		40		1⅓
Water		725	1	10
TOTALS	1	785	4	0

This is made at 25°C (78°F) and allowed to ferment for 1 hour. As an alternative an overripe bread dough can be used.

	PASTRY				FILLING			
	kg	g	lb	oz	kg	g	lb	oz
Basic dough		305		11	1	475	3	5
Soft flour		585	1	5		–		–
Butter		70		2½		110		4
Caster sugar		35		1¼		–		–
Salt		7		¼		–		–
Water		205		7½		–		–
Brown sugar		–		–		110		4
Treacle		–		–		110		4
Mixed spice		–		–		110		4
Currents		–		–	2	835	6	6
Sultanas		–		–	1	500	3	6
Mixed peel		–		–		585	1	5
Whole almonds		–		–		280		10
Rum or brandy		–		–		as required		
TOTALS	1	207	2	11½	7	115	16	0

Pastry
(1) Work the butter into the basic dough.
(2) Dissolve sugar and salt in the water, add and mix to a thin paste.
(3) Add the flour and mix until smooth and clear.

Filling
(1) Thoroughly mix the washed fruit and pour on the spirit and essence. Allow time for this to soak in – preferably overnight in a covered container.

(2) Cream the sugar and butter, beat in the eggs and add and mix in the flour and spices. Blend in the treacle.
(3) Incorporate the cake-like mixture into the tea-bread dough to make a smooth, thoroughly amalgamated mixing.
(4) Thoroughly mix in the prepared fruit and lastly the unblanched whole almonds. To produce perfectly symmetrical buns the following aids should be employed:
(a) A wooden frame into which the fruit filling is compacted prior to cutting into the appropriate sized pieces to fit the tin in which it will be ultimately baked. The size will depend upon the amount of filling used and should be capable of giving a height of approx. $6\frac{1}{2}$ cm ($2\frac{1}{2}$ in).
(b) A piece of flat wood with a handle similar to a builder's trowel, to compress the slab of fruit mixture into the frame prior to cutting.
(c) Five wooden inserts approx. 6 mm ($\frac{1}{4}$ in) thick which fit into the rectangular baking tin selected.
(d) A templet cut in the form of a cross and used to cut out the dough cover prior to it being wrapped round the fruit filling (*see* Figure 37 below).

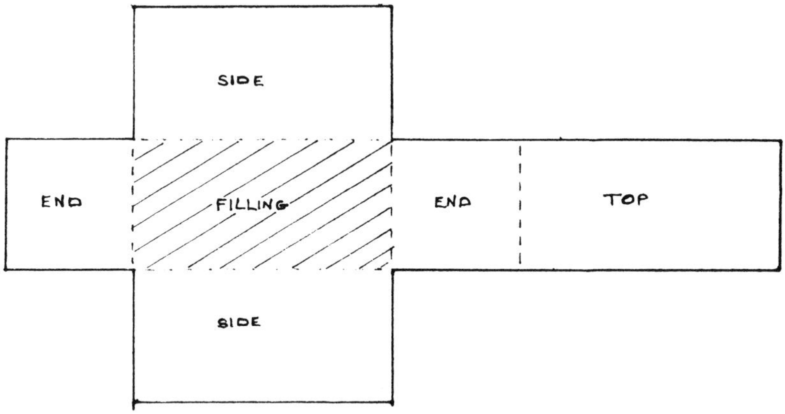

Figure 37. Shape of template used for cutting out the paste crosses for Scotch Buns. Dotted lines indicate where paste is folded. Shaded portion shows where the filling should be placed.

Detailed method of assembly

(1) Transfer the fruit filling to the frame and form a compressed slab.
(2) Divide this into the appropriate even number of pieces required by the use of a Scotch scraper.
(3) Remove the frame and transfer the preliminary cut pieces to the baking tin in which the wooden inserts have been placed. Compress the filling again, using a suitably sized piece of wood.
(4) Roll out the dough for covering to approx. 3 mm ($\frac{1}{8}$ in) in thickness and using the template, cut out in the form of crosses (*see* Figure 38 overleaf).
(5) Remove the fruit filling from the tin into which it has been compressed by turning upside down. Wash over with water.
(6) Place the rectangular or square filling into the cut dough piece and wrap it to cover the filling completely.
(7) Transfer to the baking tin into which it should now fit snugly because of the allowance given for the dough through using the wooden inserts.

Figure 38. This shows the most economical way to cut out the crosses from a sheet of paste

(8) Egg wash, dock the top and prick several times into the centre with a skewer.
(9) After proving for approx. one hour, bake at 226°C (440°F) falling to 204°C (400°F). Baking time will be approx. 40 minutes for buns weighed in at 445 g (1 lb).
(10) On removal from the oven, wash again with rum-flavoured egg wash.

FRUIT BREADS

These are made from an enriched bread dough with the addition of fruit. Some recipes use so much enrichment that they cannot justifiably be called bread; rich bun loaves would be a more appropriate description. For fruit bread the proportion of sugar should not exceed approx. 6% of the flour weight, whilst in buns it is usually approx. 18%. Tea breads fall somewhere between these two extremes.

Fruit

Almost any dried fruit may be used either on its own or in combination – sometimes with nuts.

For method of adding the fruit to the dough see page 108.

Shapes and Finishes

Although these may be baked in small tins of various shapes and sizes, simple plaited shapes may also be made. Further varieties may be made by dressing the tops with sugar and/or almond nibs before baking.

Finishes include dusting the tops with icing sugar and brushing over with water icing.

The amount of fruit can vary from as little as 25% of the flour weight to 100% for heavily fruited varieties.

The following two basic doughs are suitable for a range of fruit breads by merely adding the required fruit after mixing. The fruit should be added at the knock-back stage.

	Lightly Enriched				**Heavily Enriched**			
	kg	*g*	*lb*	*oz*	*kg*	*g*	*lb*	*oz*
Strong flour	1	000	2	4	1	000	2	4
Water		555	1	4		530	1	3
Yeast		30		1		35		1¼
Milk powder		30		1		30		1
Salt		15		½		7		¼
Sugar		30		1		70		2½
Fat		45		1½		95		3½
Egg						50		1¾
Totals	1	705	3	13	1	817	4	1¼

B.F.T. – 2 hours @ 24°C (76°F).

This can be made into a one hour dough by doubling the yeast quantity, but the bread quality may suffer slightly because the crumb will be less developed.

FRUIT BREADS USIN

(See page 1

		Currant				Sultana			
		kg	*g*	*lb*	*oz*	*kg*	*g*	*lb*	*oz*
Currants			280		10				
Sultanas							335		12
Peel (mixed cut)									
Raisins (seedless)									
Walnuts (broken)									
Dates (finely chopped)									
Wt. of fruit			280		10		335		12
Total weight with dough	*Lightly Enriched*	1	985	4	7	2	040	4	9
	Heavily Enriched	2	097	4	11¼	2	152	4	13¼

WHEATMEAL SULTANA

Any wheatmeal recipe may be used with the addition of sultanas up to 100% of the weight of flour.

However, the following slightly enriched wheatmeal sultana recipe may be preferred.

Yield – 6 loaves @ 450 g (1 lb).

	kg	*g*	*lb*	*oz*
Wheatmeal flour	1	000	2	4
Water		640	1	7
Yeast		30		1
Salt		15		½
Black treacle		20		¾
Fat		70		2½
Sultanas		860	1	15
Lemon peel		70		2½
Totals	2	705	6	1¼

B.F.T. – ¾ hour @ 24°C (76°F).

(1) Make the dough excluding the fruit and allow to ferment for half an hour.
(2) Knock back, incorporate the fruit and allow to stand a further 15 minutes.
(3) Scale and mould into warmed and greased bread tins (rectangular or round).
(4) Give a cool proof, egg wash and bake in an oven at 227°C (440°F).

THE BASIC DOUGHS
for details)

Fruit				Raisin				Raisin & Walnut				Date			
kg	*g*	*lb*	*oz*	*kg*	*g*	*lb*	*oz*	*kg*	*g*	*lb*	*oz*	*kg*	*g*	*lb*	*oz*
	165		6												
	85		3												
	30		1												
					335		12		335		12				
									140		5				
													390		14
	280		10		335		12		475	1	1		390		14
1	985	4	7	2	040	4	9	2	180	4	14	2	095	4	11
2	097	4	11¼	2	152	4	13¼	2	292	5	2¼	2	207	4	15¼

10. *Danish and Belgian Pastry*

DANISH PASTRY

Yield – 64–72 according to the varieties made

	kg	*g*	*lb*	*oz*
Flour (medium strength)	1	000	2	4
Yeast		110		4
Milk (chilled)		555	1	4
Egg		220		8
Butter or margarine		665	1	8
Cardamon spice		15		$\frac{1}{2}$
Totals	2	565	5	$12\frac{1}{2}$

Figure 39. Danish pastry varieties

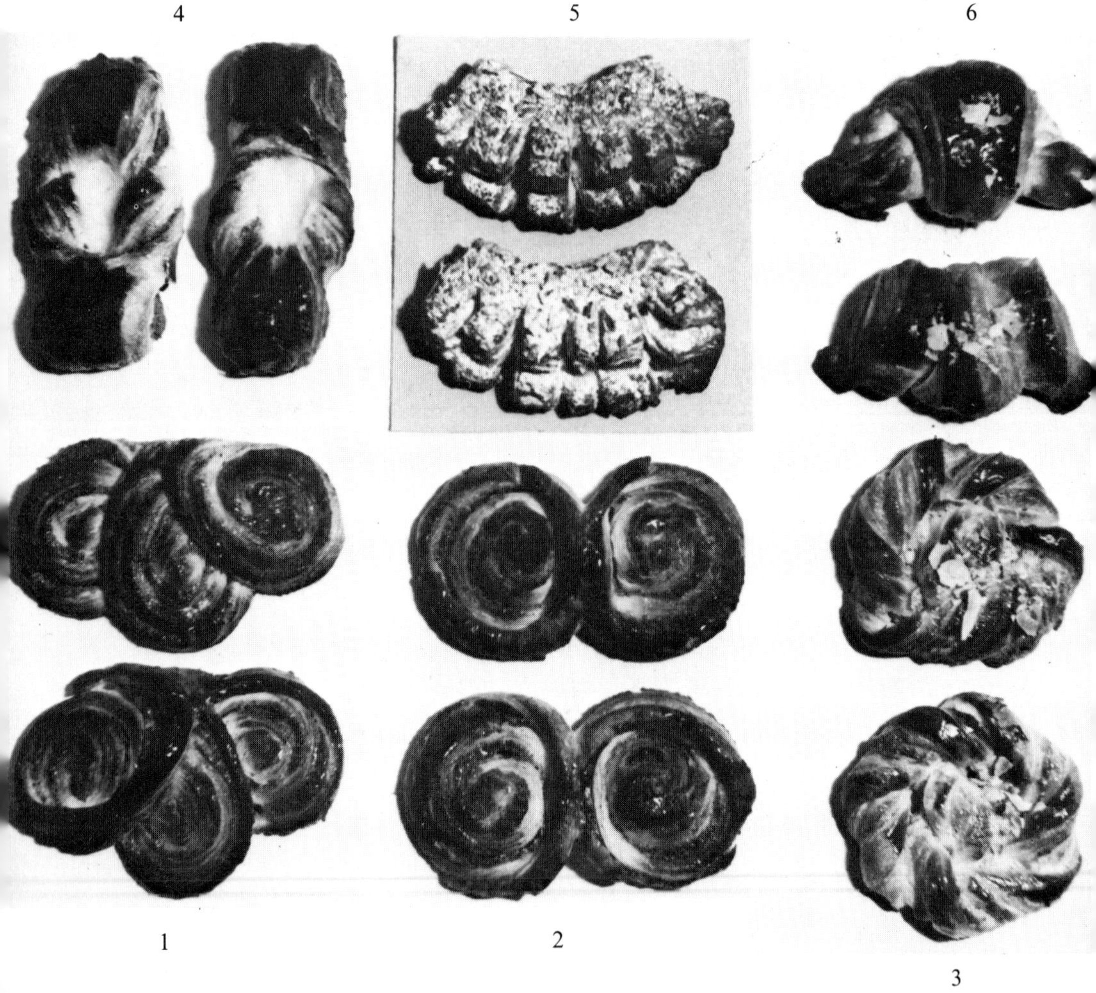

CHOICE OF INGREDIENTS

Flour

A flour of only medium strength is required. The pastry will become toughened if too strong a flour is used and it should therefore be diluted with a softer flour for these goods.

Fat

The use of a tough butter for layering will give superior results with regards flavour. If, however, margarine is to be used the following mixture of pastry and cake is preferred to the use of either on their own:

Cake Margarine 220 g (8 oz)
Pastry Margarine 445 g (16 oz)

(1) Sieve the flour and spice.
(2) Add the egg and chilled milk into which the yeast has been dispersed.
(3) Make into a dough but *do not toughen.*
(4) Roll out the dough to cover an area of approx. 40 cm × 86 cm (16 in × 30 in) and spread the butter or margarine over $\frac{2}{3}$rds of this area (lengthwise).
(5) Fold the uncoated portion of dough into the centre and fold over again so that we have three layers of dough sandwiched with two layers of fat.
(6) Roll out the dough in the opposite direction to the fold to cover the original area and again fold into three. Turn the dough through 90° and repeat the previous folding operations two more times (3 half turns as in puff pastry). Lastly give one fold over. This should be accomplished as soon as possible giving only about 10 minutes rest between turns.
(7) Give the dough a final rest for approx. 20 minutes and using the various fillings and toppings on pages 138–142 make the varieties required.
(8) Give the varieties maximum proof in a slightly humid, but cool prover.
(9) Bake in a hot oven 221°C–238°C (430°F–460°F) according to the size and type of variety chosen.
(10) When baked and whilst still hot, brush over with a rum-flavoured apricot purée and glaze with water icing.

Notes:

(1) To get the best results the dough must be cold, and therefore it is a good idea to weigh the ingredients and put them into a refrigerator beforehand.
(2) Unlike puff, Danish pastry should be very tender and this can only be achieved if the dough is not toughened during mixing. By working it off as soon as possible, proving well and baking in a hot oven for the minimum time we may be sure of getting a very flaky article.
(3) The recipe and method given has its origins in Denmark and differs quite substantially from that given from time to time in the British baking journals and information put out by various fat manufacturing firms. If puff pastry fats and margarines are used instead of butter or the mixture stated, then obviously a stronger flour will be required and more especially if a power brake is employed. This too would demand longer resting periods than those recommended here.

Retardation

The varieties made from Danish pastry may be successfully held for up to 72 hours in a retarder with a temperature 0·5°–3·5°C (33°–38°F) and a relative humidity of approx. 85%. On removal they should be allowed to recover, egg-washed, proved and baked. To prevent skin formation cover at all times with polythene sheeting.

DANISH PASTRY FILLINGS

Note All these fillings need to be of a spreading or piping consistency. Where cake crumbs are used, the milk content may need adjustment according to the type of crumbs and their moistness.

Lemon

	kg	*g*	*lb*	*oz*
Lemon curd	1	000	2	4
Cake crumbs		500	1	2
Totals	1	500	3	6

Mix together to form a cream.

Cinnamon Cream

	kg	*g*	*lb*	*oz*
Castor sugar	1	000	2	4
Butter	1	000	2	4
Ground cinnamon		7		$\frac{1}{4}$
Totals	2	007	4	$8\frac{1}{4}$

Cream together.

Apple

	kg	*g*	*lb*	*oz*
Chopped apples	1	000	2	4
Sultanas		250		9
Ground cinnamon		7		$\frac{1}{4}$
Totals	1	257	2	$13\frac{1}{4}$

Mix to a consistency which can be piped.

Almond (1)

	kg	*g*	*lb*	*oz*
Macaroon paste	1	000	2	4
Cake crumbs		780	1	12
Jap crumbs		250		9
Water		110		4
Totals	2	140	4	13

Mix the dry ingredients together and add the Macaroon paste. Bring to a piping consistency with the water.

Almond (2)

	kg	*g*	*lb*	*oz*
Ground almonds	1	000	2	4
Castor sugar	1	000	2	4
Butter	1	000	2	4
Totals	3	000	6	12

Soften the butter, mix in the other ingredients and beat to a cream. (Ground hazelnuts may also be used instead of almonds.)

Almond (3)

	kg	*g*	*lb*	*oz*
Cake crumbs	1	000	2	4
Almond paste		500	1	2
Castor sugar		500	1	2
Egg		110		4
Milk		375		13½
Totals	2	485	5	9½

Cream the almond paste with the sugar, beat in the egg, add the crumbs and milk and beat to a cream.

Fruit

	kg	*g*	*lb*	*oz*
Cake crumbs	1	000	2	4
Raisins		500	1	2
Chopped cherries		250		10
Ground cinnamon		15		½
Milk		500	1	2
Totals	2	265	5	2½

Mix together thoroughly. Adjust consistency with the milk.

Fruit and Nut

	kg	*g*	*lb*	*oz*
Cake crumbs	1	000	2	4
Sultanas		500	1	2
Currants		500	1	2
Chopped cherries		140		5
Chopped walnuts		140		5
Nib almonds		70		2½
Milk		500	1	2
Totals	2	850	6	6½

Mix the dry ingredients together and form into a spreadable paste with the milk.

Date and Nut

	kg	*g*	*lb*	*oz*
Stoned dates	1	000	2	4
Water		555	1	4
Brown sugar		220		8
Chopped walnuts		390		14
Totals	2	165	4	14

Mix the sugar, dates and water and bring to the boil. Stir in the walnuts and allow to cool.

Custard

	kg	*g*	*lb*	*oz*
Milk	1	000	2	4
Castor sugar		200		7
Cornflour		100		3½
Egg		200		7
Butter		20		¾
Vanilla flavour		as desired		
Totals	1	520	3	6¼

1. Stir together the egg, cornflour, sugar and vanilla flavour.
2. Bring the milk to the boil and pour it into the ingredients at 1, stirring to prevent formation of lumps.
3. Return the mixture to the heat, stir and cook for approx. 2 mins.
4. Lastly, stir in the butter.

Chocolate

	kg	*g*	*lb*	*oz*
Cake crumbs	1	000	2	4
Castor sugar		665	1	8
Cocoa powder		70		2½
Egg		110		4
Water		250		9
Rum flavour		as desired		
Totals	2	095	4	11½

Sieve together the dry ingredients. Add the water, egg and flavour and mix to a smooth paste.

Ginger

	kg	*g*	*lb*	*oz*
Ginger crush	1	000	2	4
Cake crumbs		250		9
Totals	1	250	2	13

Mix well together.

Pineapple

	kg	*g*	*lb*	*oz*
Pineapple crush	1	000	2	4
Castor sugar		555	1	4
Water		155		5½
Cornflour		110		4
Totals	1	820	4	1½

Make a paste of the cornflour with some of the water. Add the rest to the pineapple and sugar and bring to the boil. Add to the cornflour paste and stir vigorously. Continue the cooking for another five minutes. Allow to cool before using.

Other fillings

(*a*) *Mincemeat* – If this is too juicy to use on its own, a small quantity of cake crumbs may be added.

(*b*) Mixtures of currants, sultanas, raisins and glacé fruits, i.e. cherries, pineapple, angelica, ginger, etc.

TOPPINGS

Water Icing

	kg	*g*	*lb*	*oz*
Icing sugar	1	000	2	4
Water		150		5½
Totals	1	150	2	9½

(1) Heat the water to boiling point.
(2) Remove from the heat and whisk in the seived icing sugar to form a smooth icing.

Apricot purée

(1) Add a little water and sugar to apricot jam.
(2) Bring to the boil and strain free of lumps.
(3) Add rum to the desired flavour.
(4) Use hot.

Notes

Since jam varies so much between manufacturers it is difficult to give accurate quantities of water and sugar to add. The aim is to get a purée which will set when cold to give a glaze.

Other toppings

(a) Almonds – nibbed, strip and flaked.
(b) Sugar – granulated, castor and icing.
(c) Dried fruit.
(d) Mixtures of ground or nibbed almonds with sugar nibs.

The number of varieties are almost unlimited if combinations of fillings, toppings and shapes are considered. Basically the pastry consists of a piece of paste incorporating a filling, which may be dressed with a topping, proved, baked and usually finished with a brushing of apricot purée and icing. Varieties containing toppings such as sugar or nibbed almonds require to be baked at lower temperatures otherwise they will be too coloured. It is important to ensure that like varieties and similar sizes are baked on the same tray otherwise some will be overbaked, whilst others will be underbaked.

Varieties

Note Unless otherwise stated, i.e. "dusted with icing sugar" etc., all the varieties explained are brushed over with apricot purée and glazed with water icing immediately after baking. The icing may be flavoured with rum.

Explanation of varieties

(1) (*a*) Roll out the paste to approx. 4 mm ($\frac{3}{16}$ in) to a strip approx. 25 cm (10 in) wide.
(*b*) Spread on some cinnamon cream and cover with currants.
(*c*) Roll up like a Swiss roll.
(*d*) Cut pieces approx. 2 cm ($\frac{3}{4}$ in) wide and make two cuts with the knife almost severing the roll into 3 pieces.
Arrange the 3 pieces like a fan and lay them flat onto the baking tray (*see* Figure 39 (1)).

(2) (*a*) Proceed as (*a*) (*b*) and (*c*) above.
(*b*) Cut pieces approx. 2 cm ($\frac{3}{4}$ in) wide and make one cut with the knife almost severing the roll in 2 pieces (*see* Figure 39 (2)).
(*c*) Arrange the two pieces side by side and lay them flatly on the baking tray.

(3) (*a*) Proceed as (*a*) (*b*) and (*c*) as in previous varieties, but pin out the paste into a wider strip approx. 30 cm (12 in).
(*b*) Cut into slices approx. $1\frac{1}{4}$ cms. ($\frac{1}{2}$ in) in thickness and lay them flat onto the baking tray.

(4) (*a*) Proceed as (*a*) above.
(*b*) Cut the rolled piece in half lengthwise.
(*c*) Cut this into slices approx. 2 cm ($\frac{3}{4}$ in) wide.
(*d*) Lay pieces on the tray with the cut surface showing.

Note In the last four varieties, instead of using the cinnamon cream any of the almond fillings may be used in conjunction with dried or glacé fruits.

(5) (*a*) Roll out the paste to approx. 6 mm ($\frac{1}{4}$ in) thick.
(*b*) Spread half with a thin layer of filling, and fold over to form a sandwich.
(*c*) Roll out this piece to a width of approx. 38 cm (15 in) at an approx. thickness of 6 mm ($\frac{1}{4}$ in).
(*d*) Cut this into strips of approx. 1 cm ($\frac{3}{8}$ in).
(*e*) Starting at each end of the strip and moving the hands in alternate directions, give a twist to the strip.
(*f*) Form the twisted strip into a letter S.
(*g*) Dress the top with a sprinkling of almond and sugar nib mixture.

(6) Proceed as for the previous variety but form the twisted strip into a circle (*see* Figure 39 (3)).

(7) (*a*) Roll out the paste to approx. 6 mm ($\frac{1}{4}$ in) thick and into a strip approx. $7\frac{1}{2}$ cm (3 in) wide.
(*b*) Place some filling down the centre of the strip.
(*c*) Egg wash one edge, fold the other edge into the centre and fold over the washed edge to overlap.
(*d*) Turn the whole strip over so that the seal is underneath.
(*e*) Cut into pieces approx. $7\frac{1}{2}$ cm (3 in) long.
(*f*) Egg wash and dip the pieces into nib almonds.

(8) (*a*) Roll out the paste to approx. 6 mm ($\frac{1}{4}$ in) thick and cut into small rectangles approx. 3 cm ($1\frac{1}{4}$ in) $\times$ 9 cm ($3\frac{1}{2}$ in).
(*b*) Make a cut in the centre as in the following diagram:

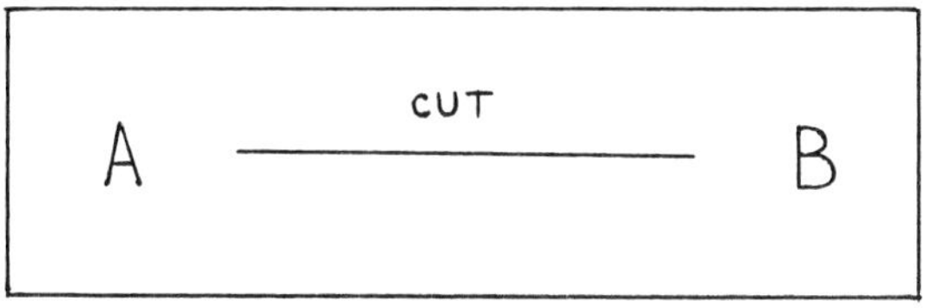

(*c*) Open the cut and thread the end B through, so that the ends are exchanged.
(*d*) When baked and glazed, pipe a bulb of custard in the centre (*see* Figure 39 (4)).

(9) (*a*) Roll out the paste to approx. 6 mm ($\frac{1}{4}$ in) and cut into squares approx. 10 cm (4 in).
(*b*) Place a portion of filling in the centre of each square and fold the corners into the centre to form a cushion.
(*c*) Egg wash and place onto the baking tray.
(*d*) When baked and glazed, a bulb of custard may be piped into the centre, alternatively a bulb of custard may be piped into the centre prior to baking.

(10) (*a*) Proceed in the same way as for No. 8 (*a*) and (*b*).
(*b*) Egg wash one edge and fold over to enclose the filling.
(*c*) Press the edge well to thoroughly seal.
(*d*) With the end of a knife, make a series of cuts along the sealed edge at about 1 cm ($\frac{3}{8}$ in) intervals.
(*e*) Egg wash and dip into nibbed almonds.
(*f*) Arrange the piece like a crescent with the cut edges on the outside (*see* Figure 39 (5)).

(11) (*a*) Roll out the paste approx. 4 mm ($\frac{3}{16}$ in) thick into a strip approx. 18 cm (7 in) wide.
(*b*) Cut this strip into triangles with a base of approx. 9 cm ($3\frac{1}{2}$ in).
(*c*) Cut a slit approx. 4 cm ($1\frac{1}{2}$ in) long near the apex (point) end of the triangle and place some filling at the base end.
(*d*) Starting at the base end, roll up the triangle.
(*e*) Form a crescent by giving the piece a twist and place onto the baking tray.
(*f*) Egg wash (*see* Figure 39(6)).

(12) (*a*) Roll out the paste to a thickness of approx. 4 mm ($\frac{3}{16}$ in).
(*b*) Cut into squares approx. $7\frac{1}{2}$ cm (3 in).
(*c*) Place a portion of filling in the centre of each.
(*d*) Egg wash two edges and fold over to form a triangle. Press edges to seal
(*e*) Egg wash and dip into nib almonds before placing onto tray.

(13) (*a*) Roll out the paste to 3 mm ($\frac{1}{8}$ in) and cut into 30 cm (12 in) strips of any convenient length.
(*b*) Spread the chocolate filling over $\frac{2}{3}$ of this and sprinkle with sultanas.
(*c*) Fold to give three layers of paste enclosing two layers of filling in a strip 10 cm (4 in) wide.
(*d*) Egg wash, sprinkle with flaked almonds and place onto a baking sheet.
(*e*) Divide into fingers using a Scotch scraper.
(*f*) Dredge with icing sugar after baking.

(14) (*a*) Roll out the paste to 3 mm ($\frac{1}{8}$ in) in thickness and cut into 10 cm (3 in) squares.
(*b*) Place a portion of mincemeat in the centre of each and egg wash the sides.
(*c*) Fold into three, sealing in the filling.
(*d*) Place onto baking sheets with the seal underneath.
(*e*) Make a cut approx. 5 cm (2 in) in length in the centre of the top layer so that the mincemeat shows through, and wash with egg.
(*f*) After baking, the centre is filled with cooked custard.

(15) (*a*) Roll out the paste to 3 mm ($\frac{1}{8}$ in) in thickness and cut into strips 9 cm ($3\frac{1}{2}$ in) wide of any convenient length.
(*b*) Pipe the apple filling in the centre of half the strips cut.
(*c*) Fold the remaining strips in half widthwise and at an angle cut at intervals along the folded edge to form a herringbone pattern as in the following diagram:

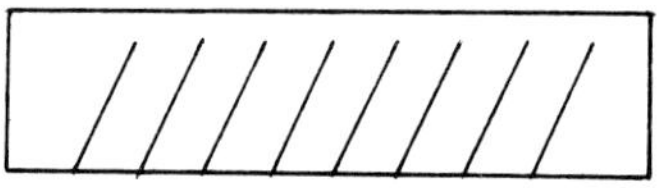
folded edge

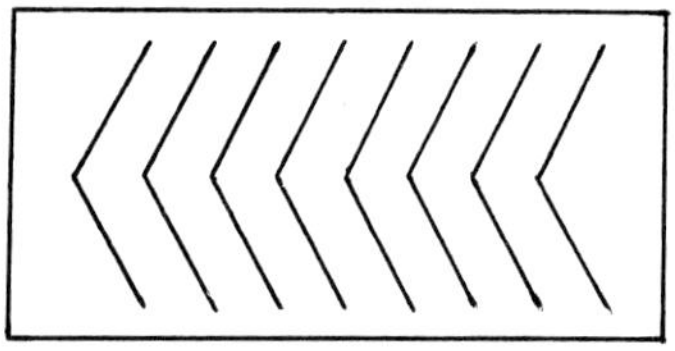
opened out

(*d*) Brush the sides of the base with egg and lay on the top piece. This is best done in the folded state, unfolding once it is in position. Press down the edges to seal.
(*e*) Produce a lattice effect by crossing over alternate cuts.
(*f*) After baking and finishing, cut into slices.

Large pastries may be made from the above varieties and sold in units sufficient for six or more persons. Often they are baked in special tins such as cottage or savarin ones.

Several units may be baked together in foil cases in the same way as fermented buns.

Belgian Pastry

This is made by taking equal parts of partly turned puff pastry and partly fermented bun dough and rolling one into the other, usually giving two half turns. The paste needs to be put into the refrigerator before working off to arrest the fermentation and prevent undue softening of the pastry fat.

Most of the Danish pastry varieties may be made using this paste, but the resultant product will not be such a nice eating article.

11. Speciality Goods

Brioche (*see* Figure 40)

Figure 40. Brioche. Placing the dough pieces into the greased tin

Yield – 72–76 @ 30 g (1 oz) approx.

	kg	*g*	*lb*	*oz*	
Strong flour	1	000	2	4	A
Milk		110		4	A
Yeast		110		4	A
Salt		30		1	A
Sugar		110		4	A
Eggs		555	1	4	A
Butter		280		10	B
Totals	2	195	4	15	

B.F.T. – 1 hour @ 24°C (76°F).

(1) Whisk the milk and into this disperse the yeast.

(2) Add the other ingredients at A and make into a well developed dough.

(3) Add the butter and beat in to form a silky, but well toughened, dough.
(4) Cover and allow to ferment for 1 hour.
(5) Lightly grease the required number of fluted patty tins.
(6) Divide the dough into approx. 30 g (1 oz) pieces.
(7) Mould first round and then to a dumb-bell shape with the small end about half the size of the other.
(8) Place the larger bulbous end of the piece in the tin. (*see* Figure 40).
(9) Keeping the small bulb suspended, make a hole with the finger in the centre of the large bulb.
(10) Allow the small bulb to rest in this impression so that the shape represents a cottage loaf.
(11) Wash carefully with egg.
(12) Prove to double the size in a cool but humid prover.
(13) Bake at 238°C (460°F) for approx. 10 minutes.
(14) Remove from the tin whilst still hot and cool on wires.

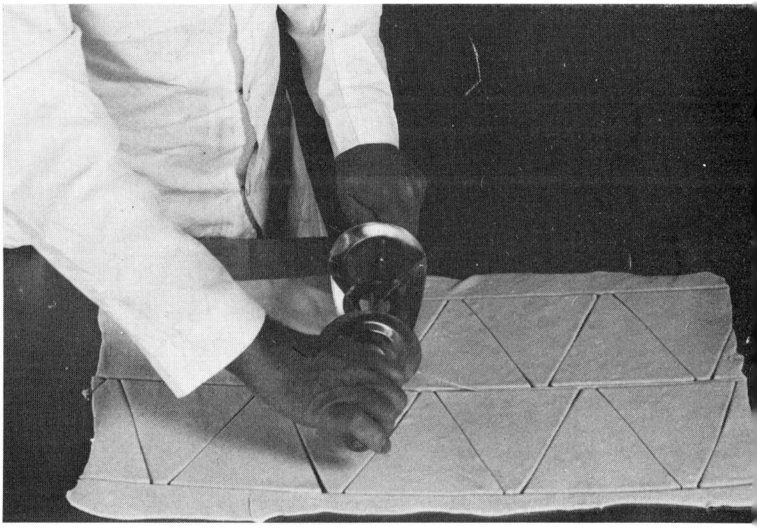

Figure 41. Method of cutting paste for croissants, using a special cutter

Croissants (*see* Figures 41 and 42)

This is a type of flaky roll which is usually served at breakfast.

Yield – 72.

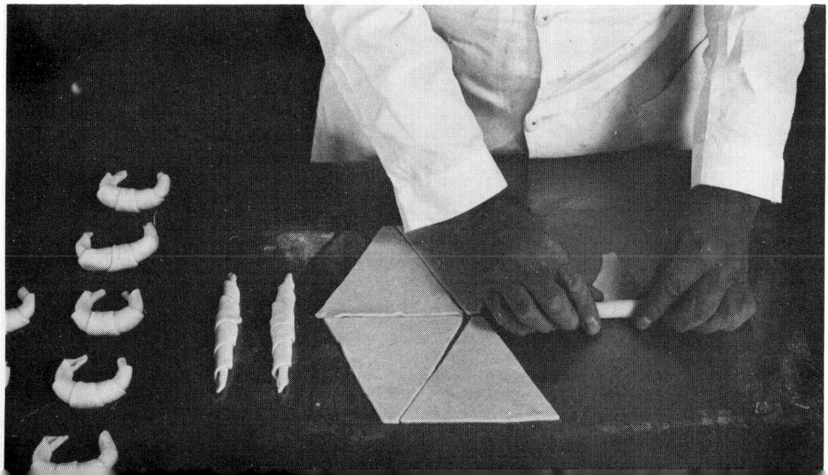

Figure 42. Croissants: showing how the paste is formed into crescent shapes before baking

	kg	g	lb	oz	
Strong flour	1	000	2	4	
Milk		665	1	8	
Yeast		45		1½	A
Sugar		85		3	
Salt		20		¾	
Tough butter or pastry margarine		445	1	0	B
Totals	2	260	5	1¼	

(1) Make a cold, well toughened dough from the ingredients at A.
(2) Allow a resting period of 30 minutes and then roll out to a rectangle approx. 12 mm (½ in) in thickness.
(3) Cover ⅔ of the dough with the butter or fat and fold as described under Danish pastry on page 137.
(4) Proceed to give three half-turns allowing a rest period between each turn.
(5) Roll the paste to a thickness of approx. 2 mm ($\frac{1}{10}$ in) and cut into strips 20–23 cm (8–9 in) wide.
(6) These are then cut into triangles having a base of 10–11 cm (4–4½ in). (*see* Figure 41).
(7) Starting at the base, lightly roll up the triangle into a crescent shape and place onto a baking sheet (*see* Figure 42).
(8) Egg wash thoroughly and prove in a cool but humid prover.
(9) Bake at 238°C (460°F) for approx. 10 minutes.

SAVARINS AND BABAS

Yield – Savarins – 9 @ 280 g (10 oz) approx.
Babas – 45 @ 55 g (2 oz) approx.

	kg	g	lb	oz	
Strong flour	1	000	2	4	
Water		110		4	
Yeast		85		3	A
Eggs		780	1	12	
Sugar		30		1	
Melted butter or margarine		500	1	2	B
Totals	2	505	5	10	

Syrup

	kg	g	lb	oz
Water		500	1	2
Sugar		250		9
Totals		750	1	11

Boil the water and sugar to form a syrup. Some of the water may be replaced by liqueurs and/or fruit juices or, alternatively flavoured.

(1) Make the ingredients at A into a well developed dough at 27°C (80°F) and allow 30 minutes B.F.T.
(2) Beat the butter or margarine at B a little at a time until it is all incorporated.

Savarins

(1) Prepare savarin moulds by giving them a liberal greasing with fat. If desired, flaked or nibbed almonds may be sprinkled on so that they adhere to the sides of the savarins during baking and so improve their appearance.
(2) Half fill the mould by piping the dough through a 6 mm ($\frac{1}{2}$ in) tube. Because of the nature of this mixing it will be necessary to sever the end with the tip of the finger once the mould is filled.
(3) Prove until dough reaches the top of the mould.
(4) Bake at 232°C (450°F) for approx. 25 minutes.
(5) Remove the mould whilst still hot.
(6) When cold, saturate with syrup. The best method of doing this is to put a measured amount into the savarin mould, replace the savarin and leave the syrup to become absorbed. It is then removed and placed upside down onto a draining wire.
(7) These goods are glazed with boiling apricot purée, filled with fruit, decorated with fresh cream and served as a sweet at a buffet or main meal.

Babas

(1) These are made from the same savarin dough, with the addition of 250 g (9 oz) currants, or any other mixture of dried fruit.
(2) Pipe into small well greased dariole moulds or deep patty tins, prove and bake at 232°C (450°F).
(3) Like savarins they are soaked in syrup and split and filled with fresh cream. For rum babas the syrup should contain rum spirit and for this to be in sufficient quantity to be detected by the consumer at least 2% based on the product weight should be added.

Stollen

This is a rich fruited dough which is made and eaten at Christmastide on the Continent, especially in Germany and Austria. A basic recipe is given here but there are many recipes in which the quantities of the enriching ingredients vary. Some contain spice (*see* Recipe overleaf).

(1) Make a ferment, cover and set aside for $\frac{1}{2}$ hour.
(2) Excluding the fruit, add the other ingredients, which should be warm, to the ferment and make into a well developed and smooth dough, the temperature of which should be not less than 24°C (76°F).
(3) Allow to ferment for $\frac{3}{4}$ hour and then carefully add the warmed fruit.
(4) After a further fifteen minutes, scale into pieces of the appropriate size.
(5) Mould in a batton shape and leave covered to recover.
(6) With a rolling pin, press down the centre of each batton lengthwise.
(7) Brush melted butter into the crease so formed, fold the halves over each other, press and place onto a warmed and greased baking sheet.
(8) Egg wash, give a cool proof and bake at 215°C (420°F) for $\frac{1}{2}$–$\frac{3}{4}$ hour, depending on size.
(9) Immediately they are baked, brush over with butter and dredge liberally with icing sugar.

Yield – 9 @ 285 g (8 oz) approx., or 8 @ 320 g (10 oz) approx.

	Ferment				Dough			
	kg	*g*	*lb*	*oz*	*kg*	*g*	*lb*	*oz*
Strong flour		110		4				
Water 38°C (100°F)		530	1	3				
Milk powder		30		1				
Yeast		55		2				
Sugar		30		1				
Salt		7		$\frac{1}{4}$				
Strong flour						890	2	0
Sugar						140		5
Egg yolks						70		$2\frac{1}{2}$
Butter						280		10
Sultanas or raisins						335		12
Cut lemon peel						85		3
Lemon zest or paste						as desired		
Total weight of dough					2	562	5	$11\frac{3}{4}$

For finishing:

	kg	*g*	*lb*	*oz*
Butter		195		7
Icing Sugar		125		$4\frac{1}{2}$

B.F.T. – Ferment $\frac{1}{2}$ hour.
Dough 1 hour.

Almond Stollen

These are the same, but with either of the following treatments:
(1) Replace the sultanas with strip almonds.
(2) Instead of brushing butter into the fold, lay in a strip of softened marzipan.

These goods should not be eaten for 24 hours. If wrapped in film they will have a shelf life of up to 1 month.

Pannetone (*see* Figure 43)

This is Italian Christmas Bread. Traditionally it is made from unleaven dough on a process lasting three days. The following Swiss recipe uses a considerably shorter process and is well recommended.

Yield – 5 @ 480 g (1 lb) approx.

	Ferment				Dough				
	kg	*g*	*lb*	*oz*	*kg*	*g*	*lb*	*oz*	
Strong flour		280		10					A
Water @ 38°C (100°F)		180		$6\frac{1}{2}$					
Yeast		35		$1\frac{1}{4}$					
Sugar						180		$6\frac{1}{2}$	B
Salt						15			
Malt						7		$\frac{1}{4}$	
Water @ 38°C (100°F)						180		$6\frac{1}{2}$	
Lemon juice						15		$\frac{1}{2}$	
Flowers of Sicily						as desired			
Strong Flour						720	1	10	C
Butter						180		$6\frac{1}{2}$	D
Egg yolks						210		$7\frac{1}{2}$	E
Sultanas						280		10	F
Orange and lemon peel						140		5	
Total Dough Weight					2	422	5	$6\frac{1}{2}$	

Butter for finishing 35 g ($1\frac{1}{4}$ oz)

B.F.T. Ferment $\frac{1}{2}$ hour
Dough 3 hours } @ 27°C (80°F).

(1) Make the ingredients (A) into a ferment, cover and set aside for half an hour.
(2) Mix the ingredients (B) and add to the ferment.
(3) Add the flour (C) and make into a well developed dough.
(4) Work the butter (D) into the dough a little at a time.
(5) Add the egg yolks (E) and make into a well developed dough.
(6) Lastly add the warmed fruit (F) and carefully incorporate.
(7) Cover and set the dough aside in a warm place for $1\frac{1}{2}$ hours.
(8) Slightly grease the hands (the dough is very soft and sticky).
(9) Weigh into suitably sized pieces and mould round with greasy hands.
(10) Allow to stand for $1\frac{1}{2}$ hours, but during this time gently mould the pieces again twice.
(11) At the end of this time gently mould to get a smooth clear skin and place the pieces into cake hoops lined with greased paper.
(12) Allow to prove until well above the level of the cake tin – approx. 1 hour.

Figure 43. Panettoni

(13) With a sharp knife, cut the top skin in the form of a cross and then place into an oven at 204°C (400°F).
(14) After five minutes, carefully withdraw from the oven, open up the cuts and place a small knob of butter in the centre of each piece.
(15) Replace in the oven to finish baking – approx. 40 minutes.

These goods should not be eaten for at least 24 hours and, if wrapped, will keep for at least two months.

Italian Easter Bread

(1) Make a batter from the ingredients (A).
(2) Disperse the yeast in the milk (B) and add to the batter (A).
(3) Add the flour (C) and make into a well developed dough.
(4) Carefully incorporate the ingredients (D), cover and set the dough aside to ferment.
(5) Scale and mould into round cobs, place onto a greased baking sheet and prove.
(6) When fully proved spread over the topping which has been well beaten. This is best done with clean hands.
(7) Dress with the strip almonds and then dust liberally with the icing sugar.
(8) Bake at 204°C (400°F) for approx. 35 minutes.
(9) On removal from the oven, dust again with icing sugar.

Yield – 8 @ 285 g (10 oz) approx. (for the dough).

	kg	*g*	*lb*	*oz*	
Castor sugar		45		1½	
Butter		85		3	
Egg		110		4	
Honey (liquid)		45		1½	A
Malt extract		15		½	
Salt		15		½	
Lemon zest or lemon paste		15		½	
Fresh Milk @ 38°C (100°F)		500	1	2	B
Yeast		45		1½	
Strong Flour	1	000	2	4	C
Sultanas		250		9	
Orange Peel		55		2	D
Strip Almonds		125		4½	
Totals	2	305	5	2½	

Topping

	kg	*g*	*lb*	*oz*
Ground almonds or hazelnuts		30		1
Castor sugar		35		1¼
Egg white		45		1½

Dusting

	g	*oz*
Icing sugar	125	4½

Dressing

	g	*oz*
Strip almonds	55	2

B.F.T. 1 hour – Knock back after 40 minutes.

12. Hotplate Goods

HOTPLATE GOODS

For these goods a purpose designed hot plate is necessary. These are heated either by gas or electricity, the latter having the advantage of a more even heat distribution and flexibility in control.

Many hotplate goods such as crumpets are now made by mass production techniques in travelling hotplates, but the methods described here refer to the smaller portable hotplates which are to be found in small bakeries.

The site of such a hotplate needs careful consideration:

(1) It should be sited in an area which is free of draught so that fermented goods will not become chilled.

(2) Good lighting is essential so that the operator can judge correctly when goods have to be turned over or when cooked and have to be removed.

(3) As there inevitably will be drips and some debris dropping onto the floor surrounding the hotplate, this should be of such material as to make cleaning easy, i.e. tiles.

(4) Adjacent shelves or racks will be helpful in storing the specialized equipment and tins required as well as quickly disposing of the baked items.

(5) Standing over a hotplate is no sinecure for the operator and good extraction of the fumes and heat should be provided.

MUFFINS AND CRUMPETS

For these a hotplate free of grease must be provided.

	Muffins				**Crumpets**			
	kg	*g*	*lb*	*oz*	*kg*	*g*	*lb*	*oz*
Strong flour @ 21°C (70°F)	1	000	2	4	1	000	2	4
Water @ 37°C (100°F)		720	1	10	1	110	2	8
Yeast		30		1		55		2
Salt		20		$\frac{2}{3}$		30		1
Sugar		10		$\frac{1}{3}$		7		$\frac{1}{4}$
Bicarbonate of Soda		—		—		3		$\frac{1}{8}$
Skimmed milk powder		30		1		—		—
Cold water		—		—		280		10
Shortening		10		$\frac{1}{3}$		—		—
Totals	1	820	4	$1\frac{1}{3}$	2	485	5	$9\frac{3}{8}$

Dough times – $1\frac{1}{2}$ hours.
Knock back muffin dough at 1 hour.

Muffins

This is a very soft dough made from warmed flour and water and requires careful handling. The dough is baked in hoops on the hotplate.

(1) After fermentation, using flour to prevent sticking, scale into the appropriate sized pieces according to the size of the hoops in which they are cooked.
(2) Mould into balls and set these onto boards liberally dusted with rice flour.
(3) Cover to prevent the formation of a skin and allow to prove.
(4) If the hoops are not siliconed they must be very slightly greased and placed onto a moderately heated hotplate to warm.
(5) Transfer the proved muffins to the hoops on the hotplate by the use of a wide chisel scraper.
(6) Bake one side and then reverse to finish cooking.
(7) Brush off the adhering rice.

Crumpets or Pikelets (*see* Figure 44 below)

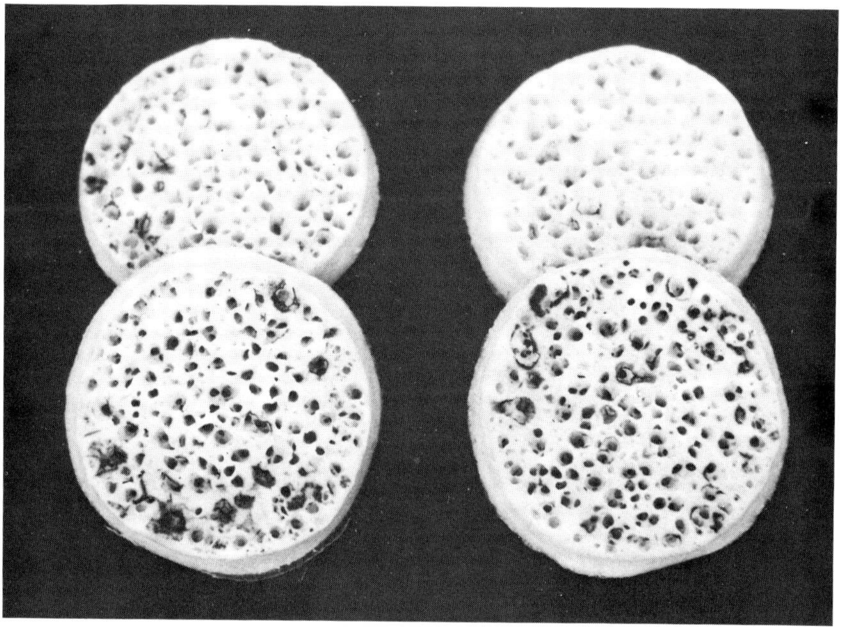

Both these names refer to the same article, but the word *crumpet* will be used here to describe the method, etc.

Making crumpets is a very skilled operation requiring a considerable knowledge of the correct condition of the fermented batter used, the temperature of the hotplate and the technique employed in their making.

Method for Making the Batter

(1) Make up the batter from the warm flour and water, excluding the soda and cold water.
(2) Give the batter a thorough beating and leave to ferment for $1\frac{1}{2}$ hours covered over.
(3) After its bulk fermentation, dissolve the soda in the cold water, add and beat into the batter.

Method for Cooking on Hotplate

(1) Ensure that the hotplate is clean, polished free of any grease and is moderately hot.
(2) Very lightly grease crumpet rings and set them onto the hotplate. Start so that space is left for one row to be turned over.
(3) Using a dipper (like a ladle) pour an even amount of the batter in each of the rings in rotation. The amount will depend upon the size of the ring and will only be properly judged by the experience gained when a finished crumpet of approx. 12 cm ($\frac{1}{2}$ in) thick is made. To prevent the batter dripping in between the hoops it is a good idea to use a shallow metal plate or sponge sandwich tin to catch the drips from the dipper. Alternatively a dropping funnel may be used (*see* Figure 45).
(4) As the batter is cooked, bubbles will appear on its surface and these will ultimately form into small holes as the mixture becomes set. When this has occurred the rings are removed and the crumpet turned over by means of a wide palette knife. By the time the last ring has been filled the first one should be almost ready for turning.
(5) Turn the crumpets as required and leave until the side with the holes are delicately coloured.
(6) Remove the cooked crumpet and prepare the hotplate and rings for the next batch.

Figure 45. Use of a hand held depositor for filling the crumpet rings. The funnel is filled with the batter which is deposited by operating a lever at the side. Notice the placing of the hoops leaving a space which is filled when the crumpets are turned over

Notes:

(1) As the cooking time is only a few minutes, several batches may be cooked from the same recipe. However, better results will be obtained from a number of small batches used in succession, than one large batch.
(2) It is usual to have enough rings for two or three batches to give time for the rings of each batch to be cooled, cleaned and greased before use.
(3) If the hotplate becomes greasy, the crumpets will not be satisfactory, so that the greasing of the rings must be minimal. Silicone-treated rings are a great advantage.
(4) The hotplate should be scraped free of any droppings and polished between each cooking.
(5) The finished crumpets may be stacked on end against each other for storing.

13. Legislation

The subjects covered in this chapter are covered by three pieces of British legislation. The Food Hygiene (General) Regulations of 1970, the Health and Safety at Work Act of 1974 and the Trade Descriptions Act of 1968. Everyone, whether employer or employee, is involved in observing and implementing the provisions of these three pieces of legislation and the courts can impose heavy penalties for offenders. The main aspects of this legislation insofar as they concern the bakery industry are given here.

THE FOOD HYGIENE (GENERAL) REGULATIONS 1970

These regulations can be conveniently divided into three parts:
PRINCIPLES AND PEOPLE.
PREMISES.
FOOD.

One part is useless without the others. It is of no use a baker or confectioner being clean in his habits or the way he handles food if the premises in which it is prepared are dirty or constructed in such a way that makes cleaning difficult. It is of no use preparing food in a hygienic way on clean premises if he/she doesn't understand the vulnerability of certain foods to grow food-poisoning bacteria if incorrectly stored. Before this latter point can be understood we need to know what causes food poisoning.

FOOD POISONING

This may be defined as an illness which can cause stomach pains, acute diarrhoea and vomitting within 1 to 36 hours after eating infected food. In extreme cases death can occur. Such foods may have become infected by:

(a) Chemicals which may have entered food accidently or during some aspect of its processing.
(b) Bacteria which may have their source in humans, animals, birds, rodents, insects, etc. The food becomes harmful either by the growth of bacteria, or the toxins they produce.

Chemicals

There are certain regulations laying down the limit in parts per million of many chemicals used in food. The trace elements usually considered to be poisonous are – arsenic, copper, lead and for these the limits are set very low. In setting these limits difficulties arise because some foods are naturally rich in minerals which would normally be regarded as harmful.

Metals used for food containers sometimes need careful consideration. For example, zinc should never be used to store moist foods, since they might become contaminated.

Contamination can also take place through machinery. For example, paraffin-contaminated bread from the use of lubricating oil in the machines.

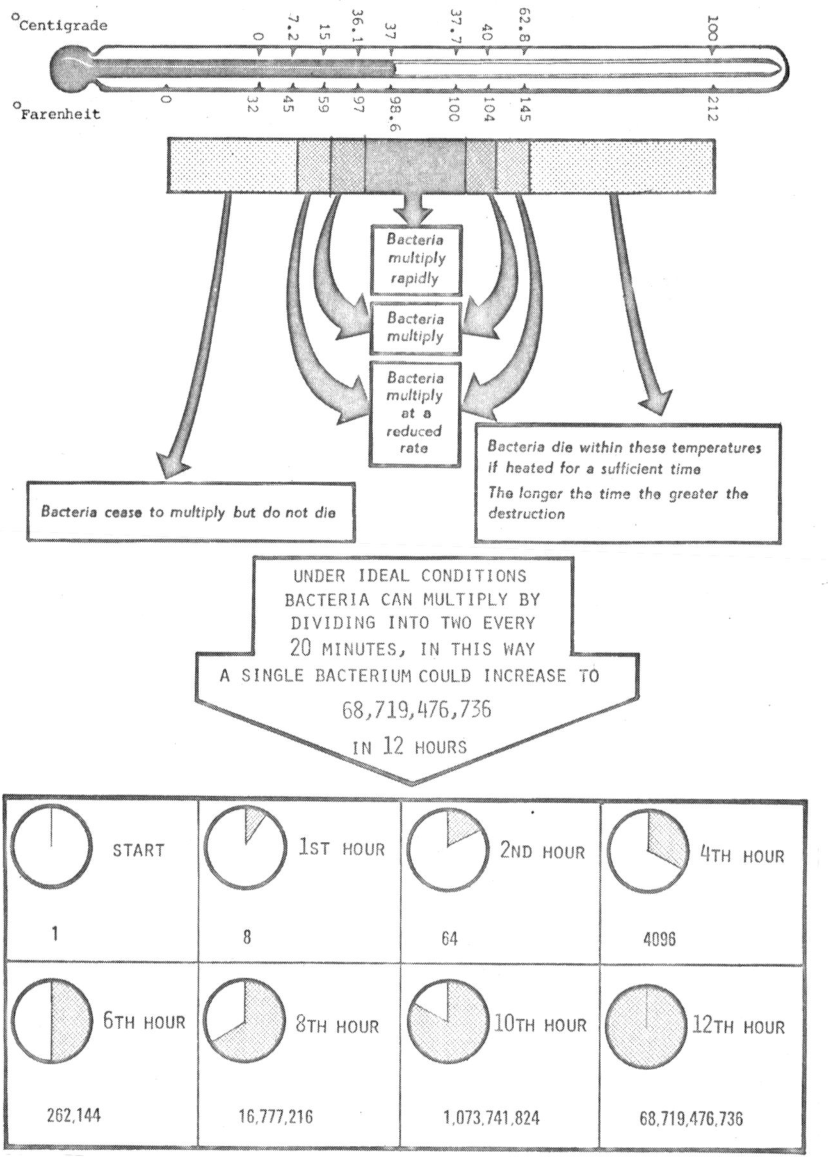

Figure 46. Effect of temperature and time on the growth of bacteria. Safe and dangerous temperatures for foodstuffs.
Reproduced by kind permission of the City of Canterbury Environmental Health Department from their handbook *Thought for Food.*

Bacteria (*see* Figure 46)

Commonly known as germs, these are so small that approx. a million could be accommodated on a pin-head. They are living organisms which reproduce themselves by simple division every 20 minutes under ideal conditions, causing one bacterium to produce several hundred million in a few hours. Not all bacteria are harmful. We use bacteria to obtain such products as cheese and yoghurt, but the harmful bacteria known as pathogens cause much suffering and many deaths each year.

The conditions required for bacteria to grow and multiply are as follows:

(1) Warmth Bacteria prefer the same temperature at which humans live, blood heat being the best temperature for rapid growth. The bacteria begin to die at temperatures above 63°C (145°F) and reproduction is slowed down when cooled to below 7°C (45°F). They cease to multiply when frozen, but become active again when thawed.

(2) Time If provided with food, warmth and water, bacteria will rapidly multiply, but before their numbers are great enough to cause illness, sufficient time is necessary.

(3) Moisture Water is necessary for the survival of bacteria and this is the reason that dried foods are incapable of supporting bacterial growth.

(4) Food Like all living forms, bacteria need nourishment and millions can survive in the smallest crack in a table. The food-poisoning bacteria thrive on the same kinds of foods enjoyed by humans especially meat, poultry, milk, eggs, coconut and jellies.

The harmful bacteria are capable of producing poisons or toxins which cause illness. Some produce the toxins outside their own cells (exotoxins) so that they mix freely with the surroundings, whilst in others the toxins are produced within the cells (endotoxins) and are not released until the organisms die.

Different symptoms of illness are produced from these two forms of toxins. If the bacteria growing in food produce exotoxin, then the food itself becomes poisonous and will give rise to signs of illness shortly after it is eaten. On the other hand, if the bacteria growing in the food produce endotoxin, symptoms of food poisoning may be delayed until they have entered and established themselves in the intestines and their numbers are such that the amount of toxin released from the dying cells is sufficient to cause illness.

It is important to realize that whilst heat might kill the bacteria, it may not destroy the toxins. Staphylococci, for example, produce toxins which are resistant to heat and may still cause illness even after the food is boiled.

The person handling food should always strive to prevent bacteria from multiplying and spreading from one place to another. This means that the source of the bacteria is eliminated and the conditions for their development are minimized.

SOURCES

Human

People who are feeling ill, suffering from diarrhoea or vomitting, who have colds, sore throats, weeping sores or are carriers should not handle food. A person who is aware that he is a carrier of an infection must inform the person responsible for the premises who in turn must then inform the Medical Officer of Health.

Animals (Domestic)

Domestic pets should never be allowed into food preparation rooms since they can be the source of infection.

Rodents

Mice and rats are able to harbour food-poisoning bacteria in their bowels and carry pathogenic organisms on their fur and in their urine. Infestation may be in the unknown quantity of food which they have contaminated either by their droppings or physical contact, as well as the small amount which is damaged. A careful watch must therefore be kept for signs of gnawing, holes, droppings and damage to packages. Baits containing poison are often recommended as a means of eradicating rodents, but since some of these poisons are not only dangerous to rodents but to humans too, expert advice should be sought from a specialist disinfestation firm or from the Environmental Health Department. Rodents can be discouraged in a number of ways:

(1) Leave no crumbs or particles of food lying around.
(2) Keep stock off the ground and rotate regularly.
(3) Refuse bins should be fitted with tightly fitting lids which should always be in place.
(4) Used paper or plastic sacks should be properly secured before being deposited in the bin area.
(5) Stored refuse should be removed from premises at frequent intervals.
(6) The building and particularly store rooms should be vermin-proofed by the provision of metal kick plates at the bottom of doors and pipe runs sealed where they pass from one room to another.

Birds

Many bakeries and other food preparation areas are plagued with birds which are attracted by food scraps left around. Windows should be screened with wire mesh to prevent their ingress. Contamination is effected by the birds' droppings.

Insects

By far the most dangerous insects are the common house fly and cockroaches.

The house fly can be eliminated by the use of insecticides, but these can only be used in dustbins, stores, etc., where food is not exposed. In food preparation areas where there is open food electrically operated fly-killers are recommended. These consist of blue light which attracts flies to an electrically charged metal grid where they become electrocuted and fall into a collecting tray underneath. All openings such as windows and doors should be screened with a fine metal gauze to prevent flying insects gaining access.

Cockroaches are known to harbour food poisoning organisms as well as those which cause dysentery, cholera and tuberculosis. This insect is nocturnal in habit and therefore if seen during the day the infestation is likely to be considerable. To breed they require warmth, moisture and food. Typical breeding grounds are in enclosed hollow spaces in walls, false ceilings, ducts, beneath service ducts or in other narrow spaces behind equipment, etc. Cockroaches emerge at night to feed on residues of food which they contaminate with their bodies or by fouling with excrement.

Because the eggs may not hatch out for several months, a number of separate treatments may be necessary. Infestation may be prevented by opening up or eliminating any concealed hollow spaces to enable all areas to be effectively cleaned. Food and grease particles must be removed at the end of each day's work. Check regularly for discarded skins and egg cases. Insecticides should not be used without expert advice, since they can be dangerous to humans and pets.

Other troublesome insect pests are silver fish and beetles and these too can be controlled with insecticides.

CLEANING

Food is liable to contamination in many ways including contact with dirty surfaces, equipment and utensils. To remove dirt and destroy bacteria requires effective cleaning and this is an essential part of any food business.

The staff employed on cleaning in a food preparation area must be regarded as just as

important as any other staff and should be properly trained. Cleaning schedules should be drawn up listing the jobs which have to be done and specifying the frequency, daily, weekly, etc. Adequate supplies of material and equipment properly suited to the various cleaning tasks should be made available. Dirty cleaning equipment spreads dirt. Wiping cloths and mops should be cleansed by boiling or with a weak bleach solution. There is a wide range of detergents, disinfectants and other cleaning materials. Any detergent used must be able to:
(1) Thoroughly wet the surface to be cleaned.
(2) Remove dirt from the surface.
(3) Hold the dirt removed in suspension.
(4) Have good rinsability.

Other factors which may need to be considered include:
(5) Bactericidal properties.
(6) Prevention of corrosion.
(7) Prevention of scale formation.
(8) Economy in use.

No single detergent meets all the above requirements, so the following terms used in connection with detergents will be of use in having to make a choice.

Soap

This is a simple detergent but lacks the good wetting action of a synthetic detergent and the powerful dissolving properties of alkalies. Soap may also form scum if used with hard water. Its use is not therefore recommended in connection with washing food equipment and utensils.

Synthetic detergents

(a) Anionic (negatively charged) detergents are the most widely manufactured synthetic detergents and the bulk of household products, both liquids and powders. Combined with a hypochlorite it gives a degree of disinfection with a reasonable cleaning action.
(b) Nonionic (neutral) detergents are unaffected by anionic reactions, i.e. they do not react with the hardness salts in water. They have less foaming properties but are very good in their ability to remove mineral oils.
(c) Cationic (positively charged) detergents are not normally used for detergent purposes, but some are employed as sterilizers as they possess marked bactericidal properties. The quaternary ammonium compounds belong to this group.

Scouring powders

These combine an abrasive and a detergent often with a bleaching agent. If too harsh they will damage the surface and in time inhibit effective cleaning. Their use is not recommended for food preparation surfaces.

Hypochlorites

These are a good disinfectant for use in food premises and if used at the correct concentration will leave little taint or smell. As they are anionic they must not be used with cationic detergents. Aluminium alloys will be corroded by a bleach solution so it must not be used with this metal.

Bleach must not be mixed with an acid water closet cleaner since it will give off poisonous fumes of chlorine gas.

General

Strong smelling disinfectants should not be used or stored in food preparation areas since the food could easily become tainted. The use of cleaning materials is a matter requiring great care and knowledge. Before purchase, the constituents of cleaning materials should be checked and only the most suitable should be purchased. If in doubt the advice of the manufacturer should be sought.

HYGIENE

The food handler is obliged under the regulations to keep clean, his hands, face and other parts of his body likely to come into contact with food, i.e. hair, scalp, and forearms when short sleeves are worn. Food poisoning bacteria can live in the human and animal gut without any outward sign of disease. These germs can be excreted with the faeces and lavatory paper will not prevent the hands becoming contaminated. Unless the hands are washed immediately after a person has used a lavatory, this will be a source of danger. Septic throats, sores, boils, stys and the nose are other sources of food poisoning bacteria. Germs can also be spread by coughing and spitting.

For these reasons hands should be kept clean and washed at frequent intervals in hot water using a detergent or soap. They should then be thoroughly rinsed and dried, preferably on a fresh paper towel. Nails should be short and if soiled cleaned by the use of a nail brush.

To summarize, hands must be washed immediately:

(1) After visiting the lavatory for whatever reason.
(2) Before handling food. If food is to be eaten without further cooking, special care is needed.
(3) After blowing the nose.
(4) After smoking.
(5) After handling raw fish, meat, poultry, vegetables or fruit.
(6) After handling refuse or swill.
(7) After handling pets.
(8) After contamination by other soiling matter.

It is recommended that liquid soap dispensers containing a liquid bactericidal soap or cream should be used. Plastic nail brushes with nylon bristles should be used as these will allow periodic disinfection by boiling or soaking in a bleach solution.

Towels should be either the paper, disposable type or a continuous roller towel system providing a portion of clean towel for each person. The following code for personal hygiene for all food handlers should be observed to prevent germs from contaminating food:

(1) Wash hands frequently and keep other parts of the body likely to come into contact with food in a clean condition. Bathe or shower daily if possible. Avoid touching the nose, ears and hair when handling food.
(2) Keep personal clothing and protective clothing and overalls clean. Hair should be covered to prevent stray hairs or dandruff from falling into food.
(3) All cuts and grazes must be completely covered with a waterproof dressing.
(4) Smoking is an offence by staff wherever food or drink is being handled. If the employee leaves the work place for a smoke, he/she must wash his/her hands before resuming work.
(5) Illness such as diarrhoea and sickness and conditions such as skin rashes must be reported to the management.
(6) Handle food for consumption with tongs and not with the fingers. Coughing or sneezing should not take place over unscreened food.
(7) Work areas must be kept clean.
(8) Sleeping in food preparation rooms is strictly forbidden.

PREMISES

The regulations make it quite clear that no food business shall be carried out on, in or at any unsanitary premises, the use of which because of the situation, the construction or condition thereof, exposes food to risk of contamination.

The place where a bakery is situated is important in this respect since it needs to be in a clean-air zone and away from sources of infection, such as rubbish dumps.

The building itself must be of a structure which is clean and able to be kept clean as follows. The room in which food is prepared must be used solely for this purpose and no other. Bedrooms and toilets must not be capable of opening directly onto a food-preparation or storage room.

Floor

This should be of smooth, impervious and non-slip durable materials which can be effectively cleaned and drained. It must relate to the type and weight of mobile equipment which has to be used. Quarry tiles with a cove between floor and wall are well recommended.

Walls

A smooth, impervious and washable surface is required from floor to ceiling. Solidly bedded, close-fitting glazed wall tiles are acceptable, except where impact damage might result. A hard-gloss, plastic based paint may also be used together with tiles to about 2 metres (6 feet). Glass fibre, laminated plastic and stainless steel panels are other suggested surfaces. Whatever material is chosen it should be light in colour. Coving at the joins between ceiling/wall and wall/floor facilitate cleaning and eliminates accumulation of dust.

Ceiling

This must be capable of being easily cleaned, resist condensation and conceal services as much as possible. It should be finished in a light colour.

Windows

Window ledges or boards should be accessible for easy cleaning. Sills should be constructed so that they cannot be used as shelves. Fly proofing is recommended. If sunlight is troublesome, sun blinds should be provided.

Doors

These should be flush fitting without panels for easy cleaning. Stainless steel kick plates should be provided on each side to prevent both damage and entry by rodents.

Equipment

Equipment which comes into contact with food must be clean and must be made of suitable materials and kept in such good order, repair and condition as to:

(*a*) Enable them to be thoroughly cleaned.
(*b*) Prevent any matter being absorbed by them as far as is reasonably practical
(*c*) Prevent any risk of contamination of the food as far as reasonably practical.

All containers intended for the storage of food must be protected from the risk of contamination. All equipment should be made mobile by the fitting of castors so that they can be moved to gain access for cleaning. Plenty of space should be given to equipment to facilitate not only their cleaning, but also the surrounding area.

Artificial Light

Fluorescent light to the Illuminating Engineering Society standard of 400 Lux on working surfaces is recommended. Background lighting above 200 Lux is desirable throughout all food premises. Light sources should be so positioned as to avoid shadows on the work surfaces by staff.

Ventilation

Artificial ventilation is usually essential in a bakery to rid the air of fumes and steam which are formed in the baking process. In some parts of the bakery humidity needs to be controlled. For example, in a dough room with exposed dough, a high humidity is desirable to prevent skin formation, whereas in cake-decorating rooms a dry atmosphere is required. Air temperature, humidity and ventilation must all be controlled to fit the specialized aspects of work in the bakery.

Water Supply and Drainage

A sufficient supply of wholesome water, both hot and cold, must be available. Cold water for drinking and use in food must be drawn directly from the main supply and not through a storage tank. Hot-water pipes should be suitably lagged with materials unaffected by heat. Water tanks should be covered, examined and cleaned regularly, i.e. at least every six months.

The drainage system must be adequate to prevent flooding and conform with the bye-laws of the local authority. Where fat or oil might get into the drainage system, grease trays should be installed and washed at regular intervals. No air intake into the drainage ventilation system must be situated in a food room.

Lavatories

These must not communicate directly with food rooms or where food equipment is cleaned. Lavatories should be kept clean and in good order, well illuminated and ventilated.

Washing

At least one deep wash basin must be provided with an adequate amount of hot and cold water, soap or another suitable detergent, nail brushes, clean towels or other suitable drying facilities.

Separate sinks with an adequate hot and cold water supply must also be provided for washing equipment and for washing food. Keeping these washing facilities separate reduces the risk of cross contamination. Hands should not therefore be washed in the sinks reserved for food and equipment, and vice versa, nor should food and equipment be washed in the same sink.

Staff Rooms

Changing facilities must be provided in rooms separate from the bakery, so that employees are attired in protective clothing before entering the work area. Often these staff rooms are equipped with showers.

Removal of Waste

Refuse and empties must not be allowed to accumulate, but be disposed of at frequent intervals to prevent the area becoming attractive to rodents and insects which is likely to bring contamination into the bakery.

Yards and Forecourts

If these are used as part of the bakery, the construction of the floors and walls should be similar with impervious surfaces which are provided with proper drainage to facilitate regular cleaning. The yard should have an alternative means of access other than the bakery.

Vehicles

Vehicles used to transport food must be so constructed that they can be easily cleaned. Hand-washing facilities should also be installed particularly in mobile shops.

Shops

Shop premises and fittings should be so arranged that the goods on sale are protected from contamination from insects, from the touch of customers and the direct rays of the sun. Wherever possible food should be covered to protect it from dust. Products containing fresh cream, meat, etc., should be kept in refrigerated counters.

FOOD

Every person handling food must take every precaution to prevent food from becoming contaminated, particularly by placing or letting it be placed in a position where the risk is apparent.

Food must not be wrapped in any paper which is not clean or which may contaminate food. Printed paper, unless exclusively designed for wrapping or containing food, must not come into contact with baked goods.

Food must not be carried in a vehicle or container along with any article from which there is a risk of contamination.

Storage Temperatures of Food (*see* Figure 47 overleaf)

Special storage temperatures are required for food consisting of meat (including poultry and game) fish, gravy or cream, or prepared from or containing any of those substances, or any egg or milk. (The term *food* in the following relates specifically to these products.)

Food brought into premises used wholly or partly for the supply of food for immediate consumption must, unless exposed for sale, be brought without avoidable delay to a temperature of not less than 63°C (145°F) or unless it is already below 10°C (50°F) must be brought to below 10°C (50°F).

Food cooked or partly cooked on the premises or food such as described in the preceding paragraph must, unless it is exposed for sale, be kept at not less than 63°C (145°F) until served for immediate consumption. If the temperature is brought or falls below 63°C (145°F) the food must be cooled below 10°C (50°F) under hygienic conditions as quickly as possible, and kept there until served or re-heated for service.

If the food falls or rises below or above the permitted temperatures during internal transit, it must be restored to the permitted levels as soon as possible.

Note *Egg* includes dried egg and *milk* includes separated or skimmed milk, dried milk, condensed milk and cream.

These regulations apply particularly to meat pies, which unless they are carefully stored are prone to develop food-poisoning bacteria of a very dangerous type.

Cross Contamination

This must be guarded against in a bakery by the thorough washing of all utensils and equipment between each operation. One source of cross contamination is the savoy bag. These should be thoroughly sterilized between one operation and another.

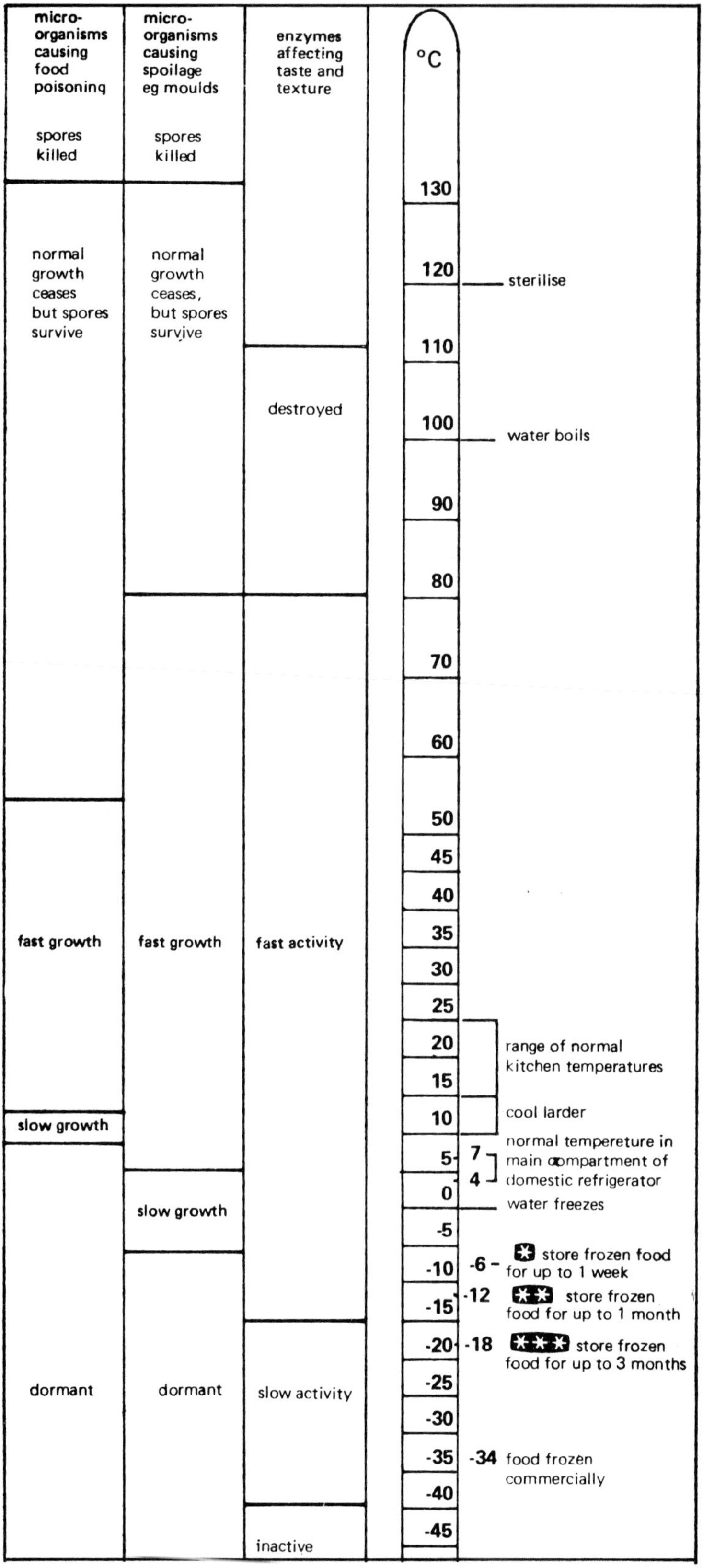

Figure 47 Food storage: a guide to temperatures. Reproduced from *Which?* magazine by courtesy of the publishers, Consumers Association

Infringement of the Law

Most shopkeepers who sell food can be prosecuted under one of two sections of the Act, with all the consequent damage to the goodwill of the business.

(1) To possess for sale or to sell food unfit for human consumption.

(2) To sell to the prejudice of the purchaser any food which is not of the nature, or not of the substance or not of the quality demanded.

Selling food unfit for human consumption is a serious charge. Baking usually renders a product safe from bacterial contamination, but there are products which are finished after baking which if improperly stored carry a risk.

Most prosecutions fall under (2) and nearly all cases taken to court concern extraneous matter found in baked goods. Most of these cases are accidently caused either by something unforeseen or by carelessness. A few may be caused with malicious intent.

An investigation must be immediately put into effect to identify the cause and produce a remedy for its non-recurrence. Sometimes the help of the Flour Milling and Baking Research Association is well recommended since they have experts on these matters.

THE HEALTH AND SAFETY AT WORK ACT 1974

This British Act of Parliament requires every employer to produce a written health and safety policy, to make the policy known to all employees, and to establish the necessary organization and arrangements to carry it out.

The following example of such a policy has been made available by the Food, Drink and Tobacco Industry Training Board in their *Guide* on this subject:

"*To all employees – Company policy on health and safety at work* The company aims to provide healthy and safe working conditions, ensure that all operations are carried out safely; co-operate with and involve all employees in meeting these two objectives.

Your board has therefore agreed to:

(*a*) Maintain necessary and up-to-date knowledge and contact with relevant outside bodies and developments on legislation, codes of practice and other technical or guidance material relating to the company's activities.

(*b*) Circulate this information within the company.

(*c*) Ensure that legal requirements are met and that steps are taken to comply with changes in these requirements.

(*d*) Train supervisory staff in accident prevention so that safe work methods are used, and systematically review retraining needs.

(*e*) Ensure that health and safety factors are taken fully into account when new methods, processes or premises are being planned, or when changes in existing products or production methods are considered.

(*f*) Give new employees and workers redeployed to new jobs basic training in safety and the availability of medical facilities and authorized first-aid personnel.

The board of directors has agreed this policy and will, with the co-operation of the safety committee, regularly check how well it is working.

Procedures To implement this policy the company will:

(1) Write the responsibility for safety and, where appropriate, health into all job descriptions and training sheets.

(2) Name a Safety Officer on each site who will advise on safety matters and investigate unsafe conditions/practices and accidents with management.

(3) Maintain a safety committee on each site to make recommendations on health and safety.
(4) Publish general safety rules (in company handbook).
(5) Ensure that all employees, including specialist operators, are trained in their duties/responsibilities.
(6) Maintain adequate provisions for fire prevention and fire fighting in consultation with the local fire service.
(7) Maintain an emergency evacuation system.
(8) Maintain medical or first-aid facilities on each site.
(9) Ensure that *good housekeeping* standards are met.
(10) Operate a preventive maintenance system.
(11) Provide necessary protective clothing.
(12) Keep adequate safety reporting follow-up procedures and statistics.
(13) Ensure that this policy covers everyone including visitors, customers and contractors."

Responsibility of employers

Section 2 (2) (e) of the new Act imposes on every employer the following duty:

"The provision and maintenance of a working environment for his employees, that is, so far as is reasonably practicable, safe and without risks to health but adequate as regards facilities and arrangements for their welfare at work."

His first duty is to "ensure as far as reasonably practicable, the health, safety and welfare at work of all his employees."

As far as is reasonably practicable the matters to which this duty extends are:

(*a*) Provisions of maintenance of plant and systems of work that are safe and without risks to health.

(*b*) Arrangements for ensuring safety and absence of risks to health in connection with the use, handling, storage and transport of articles and substances.

(*c*) The provision of such information, instruction, training and supervision that is necessary to ensure the health and safety at work of his/hers employees.

(*d*) Any place of work, under the employer's control must be maintained in a condition that is safe, without risks to health, and means of access to and egress from it that are safe and without risks to health must be provided and maintained.

The term in the Act "so far as is reasonably practical" is thought by some people to have weakened the Act. It should be realized, however, that if a matter goes before the courts, tha Act specifically places the duty on the accused if he avails himself of this defence, to prove that it was not reasonably practicable to do more than was done to meet the duty requirements, or that there were no better practical means than were in fact used.

Section 2 (3) of the Act demands that employers – "prepare and, as often as may be appropriate, revise a written statement of this general policy of health and safety at work of his employees of the time being in force for carrying out that policy and to bring the statement and any revision of it to the notice of all employees."

The Act exempts those small employers employing less than five from this requirement. Basically this statement will consist of three sections:

(1) Declaration of intent giving the names of the senior member of the management with overall responsibility for health and safety.
(2) A section dealing with the organization of health and safety including a definition of the line management's responsibilities, the function of specialist allocation of resources and role of safety representatives or safety committees.
(3) A final section dealing with the arrangements through which the policy is to be implemented.

It is the intention of the Act that individual employees shall also be accountable for their actions or omissions, especially those which result in contravention of the Law.

The duties placed upon employees include:

(1) Taking reasonable care for the health and safety of himself and other persons who may be affected by his acts or omissions at work.
(2) In relation to any duty or requirement imposed on his employer or any other person, to co-operate with that person insofar as is necessary to enable the duty or requirements to be performed or complied with.
(3) No person shall intentionally or recklessly interfere with, or misuse anything provided in the interests of health, safety or welfare in pursuance of any relevant statutory provision.

It is not the intention of this chapter to give more than a brief example of some aspects of this important piece of legislation in so far as it can be applied to the Baking Industry. If further information is required, the reader is advised either to refer to the Act itself or read the references at the end of this chapter.

SAFETY HAZARDS IN A BAKERY

Floors, Passages and Stairs

Section 28 of the Factories Act 1961 simply states:

"All floors, steps, passages and gangways shall be of sound construction and properly maintained and shall, as is reasonably practicable, be kept free from any obstruction and from any substance likely to cause persons to slip."

It also requires that openings in floors be fenced off (e.g. dough hoppers) and hand rails be provided on stairs.

Safe Access

Provision must be made for employees to have a safe access to do jobs. Examples are access to stores or roofs to do maintenance work, etc.

Ladders

These may be provided to gain access to materials which have been stacked at a height in a store. Ladders should be adequately secured at or near the top to prevent sideways motion and at the base to prevent slip. They should have a firm footing, be soundly constructed and regularly inspected to ensure that the uprights and rungs are properly maintained.

Machinery

Many machines used in bakeries are highly dangerous and great care is often required by the operator in the maintenance of health and safety. All machines which are dangerous should be guarded in one of the following ways:

Fixed Guards

This type should be provided whenever possible, but to be effective it should:

(1) Prevent all access to the dangerous part from all angles and during all operations;
(2) be of robust construction and be firmly and securely installed;
(3) be carefully designed so as not to impede efficient operation of the machine;
(4) be arranged so that it does not have to be removed for routine maintenance and lubrication.

Interlocking Guards

Where fixed guards are impracticable (e.g. on open-pan dough mixers) interlocking guards should be fitted, but it should ensure that:

(1) It has to be closed in such a way as to prohibit access to the dangerous part before the machine can be operated.

(2) It remains in the closed position until the dangerous part is at rest.

Automatic and Trip Guards

These guards should be used if neither the fixed nor the interlocking guards are practicable (e.g. pastry rollers, tart machines, etc.) It should ensure that if the position of the operator is such that he is about to be injured by the dangerous part then:

(1) If an automatic guard, it will physically remove the operator from the danger zone, *or*

(2) If a trip guard, it will cause the machine to stop or to reverse its motion before the operator can reach the dangerous part.

Photo-electric Guards

For certain machines where the use of fixed guards are impracticable, these guards are particularly suitable. A number of light beams across the opening between the operator and the dangerous parts, with associated photo-electric detectors ensure that:

(1) As long as the light curtain is interrupted the dangerous parts of the machine cannot move.

(2) If the light beams are interrupted while the dangerous parts are moving, the machine will stop.

Controls

A machine is safe only if the controls, whether electrical or mechanical are easily accessible, clearly identifiable and protected against accidental operation.

Electrical Equipment

Some type of electrical equipment is used in every factory and is subject to the Electricity (Factories Act) Special Regulations. The great majority of the electrical accidents and fatalities happen at ordinary mains voltages (around 240 volts AC).

Electricity should be treated with respect. Attention should be paid to the following points:

(1) Only competent electricians should install equipment.

(2) Regular inspection and servicing should be carried out by competent electricians. Workers should be instructed to report promptly any signs of a fault or defect.

(3) Defective equipment should be taken out of use and completely disconnected from the electrical supply until it is repaired.

(4) Electrical equipment on which repairs or adjustments are to be made should be disconnected from the supply before such work is started. The importance of this rule should be stressed to each electrician and maintenance worker.

(5) Temporary wiring should be avoided. Where it is necessary, it should be of a safe standard; should be inspected and, if necessary, repaired at frequent intervals; it should be replaced as soon as possible by a permanent installation. Efficient earth connections must be provided however short a time the temporary installation is expected to be in use.

(6) Loading of circuits should be carefully supervised. Overloading increases the risk of fire. All circuits must be protected by fuses or circuit-breakers of correct rating.

(7) Immediate artificial respiration (and external cardiac resuscitation if necessary) is essential if a person unconscious from electric shock is to be revived. Workers

should know how to apply artificial respiration, and an electric shock placard giving instructions on treatment should be on permanent display.

(8) Portable equipment which is gripped firmly by the operator accounts for a high proportion of all electrical accidents and fatalities and needs particular care:
   (*a*) Earthing is the most important precaution. No portable tool, except a certified double-insulated or all-insulated tool, should ever be used unless it is properly earthed, and the efficiency of the earth and of the insulation should be regularly tested by a competent electrician.
   (*b*) Cables and plugs should be carefully looked after, particularly at the place where the cable enters the tool and the plug, as wear in these places can cause a serious accident. At both places efficient cable or *cord* grips should be used.
   (*c*) Flexible cables should be positioned so that heavy equipment or materials will not come into contact with them.
   (*d*) Wherever possible, low-voltage tools – 110 volts or 50 volts in damp conditions – should be used. The low voltage supply should be obtained from a single phase transformer which has the centre point of the secondary winding earthed, or a three-phase transformer which has the star point of the secondary windings earthed.

(9) Electrical leads should have wires of the standard colours: brown for live; blue for neutral; green and yellow for earth. Leads of other colours on old or imported equipment should wherever possible be replaced or at least see that all concerned know about them.

Hand Tools

Although a baker or confectioner does not use many hand tools, some that are employed are dangerous if used incorrectly, e.g. knives. These should always be kept sharp, held correctly and with the cutting action away from the user.

Movement of Materials

The largest single cause of factory accidents is the movement and manipulation of goods and materials by hand. Many of these accidents, though simple in origin, result in painful disablement and long periods away from work.

The general principles of lifting and carrying are elementary but are not sufficiently widely appreciated. Correct methods are, however, quickly and easily taught and can be demonstrated without elaborate equipment. Instruction in good practice could reduce the number of accidents considerably. Accidents could also be reduced by the introduction of suitable mechanical devices wherever possible.

The Factories Act (section 72) states that no one must be employed to lift, carry or move any load so heavy as to be likely to cause injury.

Fire precautions

Asphyxiation by the products of combustion – not burning – is the main cause of death in conflagrations.

Adequate means of escape in case of fire are a vital necessity in all premises. Except for some small premises all factories and offices must have a certificate from the Fire Authority that they are provided with such means of escape in case of fire as may reasonably be required.

The effective enclosure of staircases is essential. If a stairway is not effectively separated from the workrooms on the various floors, it will be quickly filled by smoke in the early stages of a fire. People deterred from using it to escape are then liable to be trapped. Access from landings to the various rooms should be by self-closing doors and effective action should be taken to prevent such doors from being wedged open.

Fire exits must be clearly marked, and in particular such means of escape as stairways and landings kept free from obstruction.

Other precautions to reduce personal hazard are:

(1) A fire-alarm system clearly audible in every part of the factory, including washrooms and lavatories, and easily heard above machinery noise. Operating points should be sited to allow people to turn them on as they leave the building.

(2) A proper fire routine which is clearly understood by all workers, and includes fire fighting and evacuation. New workers should receive early instructions. Fire drills are essential in many cases.

(3) Adequate first-aid fire-fighting equipment and training of sufficient employees in its use. Discretion is essential in deciding the length to which first-aid fire fighting is carried and, as a general rule, it should cease and the premises should be entirely evacuated before smoke or fire threatens the means of escape.

The local Fire Authority should be consulted for advice on fire precautions or fire prevention.

HEALTH

Various sections of the Act relate to the conditions for the health and welfare of the employees. Some of these are already embodied in the Food Hygiene (General) regulations. Others which relate to bakeries are as follows:

(1) *Cleanliness* Every factory must be kept clean. In particular accumulations of dirt and refuse must be removed daily from floors and benches, the floor of every workroom must be cleaned at least once a week, and all inside walls, partitions and ceilings must (a) if they have a smooth impervious surface be washed with hot water and soap or cleaned by other approved methods every 14 months, or (b) if kept painted in a prescribed manner or varnished, be repainted or revarnished at prescribed intervals (of not more than 7 years) and washed with hot water, etc. every 14 months, or (c) in other cases be whitewashed or colourwashed every 14 months. (Certain factories and parts of factories are excepted from these provisions because of the nature of the work carried on.) The prescribed particulars must be entered in the General Register (Section 1).

(2) *Overcrowding* A factory must not be overcrowded. There must be in each workroom at least 400 cubic feet of space for every person employed, not counting space more than 14 feet from the floor (Section 2).

(3) *Temperature* A reasonable temperature must be maintained in each workroom by non-injurious methods. In rooms in which a substantial proportion of the work is done sitting, and does not involve serious physical effort, the temperature must not be less than 15·5°C (60°F) after the first hour, and at least one thermometer must be provided in a suitable position (Section 3).

(4) *Ventilation* Adequate ventilation of workrooms must be secured by the circulation of fresh air.

Working Clothes

The Food Hygiene Regulations lay down that employees engaged in food production must be properly attired in protective clothing, but in the interest of safety there are other considerations as follows:

(*a*) Jewellery, e.g. watches, key rings, rings, necklaces, etc., should never be worn. If the part of the body containing these adornments happens to get caught in a machine, the user could loose a finger, hand, arm, etc. Scarves, ties or loose clothing should also never be worn near machinery.

(*b*) Hair should always be covered and preferably kept short for the same reason. One has only to see photographs of the unfortunate victims who have been scalped by their hair being caught in a machine to appreciate the danger.
(*c*) Suitable footwear is also advisable. Shoes should be sufficiently robust to protect the foot from accidental knocks.

Fire

Measures to reduce the risk of fire breaking out or to prevent it from spreading will reduce both personal hazards and material losses. For example:
(*a*) Rubbish and process waste that could be ignited by a discarded cigarette should not be allowed to accumulate.
(*b*) The fire risk of all materials and processes (including methods of heating) should be known. Sources of ignition should be protected or eliminated.
(*c*) Machinery and plant, particularly electrical equipment, should be regularly inspected.
(*d*) Combustible stocks should be tidily and compactly stacked.
(*e*) Openings in floors and around ducts, or process plant and machinery, should be closed, as they provide direct vertical paths for smoke or flame. Hoistway enclosures should be of a suitable standard of fire resistance.
(*f*) So far as possible, manufacturing areas should be subdivided by walls of fire-resisting construction.

All practicable measures must be taken to protect workers against inhalation of dust, fumes or other impurities likely to be injurious or offensive, and local exhaust ventilation must be provided and maintained where practicable (Sections 4 and 63).

(5) *Lighting* There must be sufficient and suitable lighting in every part of the factory in which persons are working or passing (Section 5).

(6) *Drainage of floors* Where wet processes are carried on, adequate means for draining the floor must be provided (Section 6).

(7) *Sanitary accommodation* Sufficient and suitable sanitary conveniences, separate for each sex, must be provided subject to conformity with standards prescribed by regulations. The conveniences must be maintained and kept clean and effective provision must be made for lighting them (Section 7).

(8) *Underground rooms* No work is to be carried on in any underground room (unless used only for storage or other specially excepted purpose) if the District Inspector certifies that it is unsuitable as regards height, light or ventilation, or on any hygienic ground, or because the means of escape in case of fire are inadequate. Notice must be given to the District Inspector before an underground room is used as a workroom in a factory, if it was not so used on 1st July 1938 (Section 69).

(9) *Lifting excessive weights* No one must be employed to lift, carry or move any load so heavy as to be likely to cause injury (Section 72).

First Aid

In every factory there must be provided a first-aid box or cupboard of the prescribed standard, containing nothing except first-aid requisites. Where more than 150 persons are employed at one time an additional box or cupboard for every additional 150 persons or fraction of that number is required. Each box or cupboard must be placed in the charge of a responsible person who, in the case of a factory where more than fifty persons are employed, must satisfy prescribed conditions as to training in first-aid treatment. The responsible person must always be readily available during working hours and a notice must be affixed in every workroom stating the name of the person in charge of the box or cupboard provided in respect of that room (Section 61).

TRADE DESCRIPTIONS ACT, 1968

The Trade Descriptions Act, 1968, seeks to ensure that traders tell the truth about goods, prices and services.

Specific

The Act does three things:

(1) *Untrue statements* The Act lays down that if traders make certain kinds of untrue statements about the goods, prices and services they are offering, they will be liable to a fine or imprisonment.

(2) *Enforcement* The Act makes local weights and measures authorities responsible for seeing that this law is observed and gives them powers to help them do this job.

(3) *Dept. of Trade Powers* The Act gives the Dept. of Trade the power to give particular meanings to words used in the course of business, powers to require that goods be marked with – or accompanied by – information, and power to require that advertisements contain, or refer to, information.

Untrue Statements – Goods

It is an offence for any person in the course of trade or business to:

(*a*) Apply a false trade description to any goods

or

(*b*) To supply or offer to supply any goods to which a false trade description has been applied.

(A person exposing goods for supply or having goods in his possession for supply shall be deemed to offer to supply them.)

This offence raises three questions:

(1) What is a *trade description*?
(2) What does *false* mean?
(3) What does *applied* mean?

What is a Trade Description?

A trade description is an indication, direct or indirect, by whatever means given, of any of the following matters with respect to the goods:

(*a*) Quantity, size or gauge. Quantity includes length, width, height, area, volume, capacity, weight and number.

(*b*) Method of manufacture, production, processing or reconditioning, e.g. *home-made.*

(*c*) Composition, e.g. 10% *butter.*

(*d*) Fitness for purpose, strength, performance, behaviour or accuracy.

(*e*) Any physical characteristics other than those mentioned.

(*f*) Testing by any person and the results thereof.

(*g*) Approval by any person or conformity with a type approved by any person.

(*h*) Place or date of manufacture, production, processing or reconditioning.

(*i*) Person by whom manufactured, produced, processed or reconditioned.

(*j*) Other history, including previous ownership or use.

What does False mean?

The indication must be false to a material degree before an offence is committed. It is not enough for it just to contain an insignificant inaccuracy. It must be false enough to matter. Misleading indications will be treated as false indications, e.g. a cake described as weighing 480 g when in fact it only weighed 479 g would not be false in a material degree, but even a slight inaccuracy in describing precision goods could be false to a material degree, because the inaccuracy could be critical.

What does Applied mean?

(1) A person applies a trade description to the goods if he:

(*a*) *Affixes* or *annexes it to* or *in any manner marks it on* or *incorporates* it with the goods themselves.

(*b*) *Affixes* or *annexes it to* or *in any manner marks it on* or *incorporates it* with anything *in*, *on* or *with which* the goods are supplied.

(*c*) If he places the goods *in*, *on* or *with anything* which the trade description has been affixed or annexed to or marked on or incorporated with.

(*d*) Uses the trade description in any manner likely to be taken as referring to the goods.

(2) An *oral statement* may amount to the use of a trade description.

(3) A trade description will be deemed to be applied to the goods by the seller when he supplies goods to a customer who, in asking for them, has used a trade description.

(4) A trade description may be applied to an advertisement. Advertisement includes catalogue, circular, price list.

Summary

The trade description must be applied to the goods whether in writing or by means of an illustration, symbol or other marking on the goods.

Bakers can be fined under this Act for supplying bread:

(1) Ordinary wheatmeal bread baked in a proprietary tin with a branded name embossed in the side.

or

(2) Another type of proprietary bread baked in a plain tin and sold as ordinary bread.

Recommended price means:

(1) The price recommended by the manufacturer or producer,

and

(2) That it is the price recommended for supply by retail in that area.

It is an offence to give by whatever means a *false indication* that:

(1) The price is equal to a previous price.

(2) The price is less than a previous price.

(3) The price is less than a previous price by a particular amount.

An indication that goods were previously offered at a higher price or at a particular price means:

(1) The price was that of the seller *unless otherwise stated.*

(2) The price indicated was in use during the past six months for a continuous period of not less than 28 days *unless otherwise stated.*

Any indication, however given, that the price of the goods is less than that at which, in fact, they are being offered, shall be an offence.

Summary

Certain kinds of false indications about the price of goods, for example, false comparisions between a trader's current price and his own previous price; false comparisons with a recommended price; false indications that the price is less than it really is; are offences.

A trader who claims that he has reduced his price must have charged the old price for at least 28 days in the last six months *unless he expressly indicates the contrary.*

A trader who compares his price with a recommended price must compare it with a price which a manufacturer has recommended generally for retail sale in that area where the goods were offered *unless* he makes it clear that he is referring to a different kind of recommended price.

Untrue Statements – Services

It is an offence to make false statements, either knowingly or recklessly, about services, accommodation and facilities.

The statements covered by the Act are as follows:

(*a*) The provision of services, accommodation or facilities.

(*b*) The nature of services, accommodation or facilities.

(*c*) The time at which services, accommodation or facilities are provided.

(*d*) How services, accommodation or facilities are provided.

(*e*) By whom services, accommodation or facilities are provided.

(*f*) The examination, approval or evaluation by any person of services, accommodation or facilities.

(*g*) Where accommodation is provided or what amenities it has.

The Dept. of Trade has power to define terms used in respect of the above.

Enforcement

It is the duty of every local weights and measures authority to enforce, within their area, the provisions of the Trade Descriptions Act. To enable the authorities to carry out these duties the Act confers certain powers.

These powers are:

(1) The power to make test purchases.

(2) The power to enter premises and inspect and seize goods and documents.

It is an offence to obstruct an officer in carrying out his duties.

The Act requires the authorities to give notice if having made a test purchase, they intend to investigate further with a view to prosecution.

Department of Trade powers

The Act gives power to the Department of Trade to require:

(1) That expressions used in the course of trade or business which are used as trade descriptions or part of trade descriptions shall be given a definite meaning.

(2) That goods be marked with or be accompanied by information or instructions relating to the goods. The Department may also decide the form and manner in which the information is to be given.

(3) That advertisements of goods include specific information about them or refer to information about them.

Defences

It will be a defence for a person charged with an offence under this Act to prove:

(1) That the commission of the offence was due to:

(*a*) a mistake,

or (*b*) reliance on information supplied to him,

or (*c*) act or default of another person,

or (*d*) an accident,

or (*e*) some other cause beyond his control.

Provided he can show:

(2) (i) that he took all reasonable precautions instructing staff

and

(ii) exercised all due diligence (seeing that instructions are carried out) to avoid the commission of an offence either by himself or any person under his control.

Penalties

The present penalties for offences are:

(1) If tried before a jury, an unlimited fine or imprisonment up to 2 years, or both.

(2) If tried by a Lay Bench or a Magistrate, a fine not exceeding £1,000.

FOOD LEGISLATION

The amount of legislation which concerns food is to say the least exhaustive and sometimes confusing. Moreover this situation is likely to get worse as we are to now comply with the legislation imposed by the E.E.C.

Rather than compile a section of this book on legislation in so far as it affects the baking industry, the author felt that any relevant legislation was best explained in the same chapter as the goods for which the legislation was framed.

REFERENCES

Food Hygiene

(1) The Food Hygiene (General) Regulations 1970.

(1) *Thought For Food* by the City of Centerbury Environmental Health Dept.

(2) *Up To Date Breadmaking* by W. J. Fance & B. H. Wagg.

(3) *Your Guide to the Food Hygiene (General) Regulations* 1970, by The Health Education Council.

Health and Safety

(1) The Health and Safety at Work Act of 1974.

(1) *Training for Health and Safety at Work* by the Food, Drink and Tobacco Industrial Training Board.

(2) *Health and Safety at Work* H.M.S.O. Booklet No. 35.

(3) *The Thinking behind the Health and Safety Law*–Paper given to the British Chapter, American Society of Bakery Engineers by J. J. Davey, B.A.

Trade Description Act

(1) The Trade Description Act 1968.

(2) *Guide to the Trade Description Act*–Notes issued by the Grocers' Institute.

Glossary

Definitions of technical terms used in bread, fermented goods and flour confectionery.

Absorption – The soaking up of one substance into another. Usually refers to the amount of water which may be used to form a dough of workable consistency from a particular flour.

Activated Dough Development (ADD). – A dough-making process in which the development of the dough is achieved entirely by chemicals without any bulk fermentation or excessive mechanical manipulation. This can be achieved by small amounts of cysteine (35 ppm.), ascorbic acid (50 ppm.) and bromate (25 ppm.).

Additives – Any substance other than the basic raw materials of the recipe, added to a foodstuff in small quantities to improve the ease of manufacture, its quality or shelf-life. The use of most of these is strictly controlled by Law.

Adsorption – The state of holding water, i.e. the concentration of a substance on a surface.

Aerate – To charge with gas, e.g. carbon-dioxide, air, etc.

Aerograph – This consists of an air-brush which is connected by braided hose to an air supply of approx. 30 lb. per square inch pressure, 200 kPa, supplied by a compressor. It is a pen-shaped instrument which is used to direct a finely atomized spray onto icing, etc., for decorative purposes. The operator can control both the amount of colour and air pressure by a small lever on the instrument so that colour may be applied as delicately or intensely as required.

Albumen – A protein found in egg whites, to which this term is generally applied in a bakery.

All-in-Process – Mixture of all the ingredients without any preliminary stages.

Arabic – The dried gum from the acacia tree.

Ash Content – Mineral matters in foodstuffs.

Baffle Plate – A movable plate to fend off excessive heat in an oven.

Bag – Before metrication –

(a) 140 lb ($63\frac{1}{2}$ kg) of flour. Two bags equal 1 sack – 280 lb (127 kg).

(b) A cone made from paper, cloth, plastic or nylon used for icing and piping.

Bain Marie

(a) A water bath into which another vessel can stand in order that its contents may be heated.

(b) A double saucepan, i.e. porriger, in which the inner vessel is heated by the water from the outer vessel, so that direct heat is avoided.

Bake Off – Baking cakes, etc., after they have been prepared for the oven.

Baking Powder – An edible chemical or mixture of chemicals which, in the presence of moisture and heat, will generate gas (e.g. carbon dioxide) which will aerate baked goods.

Baking Sheet – Metal sheet on which bread and confectionery goods are baked. They usually have three turned-up sides and an open end, but may also be obtained with four turned-up sides.

Barm – A liquid in which yeast has been grown. Since it contains a mixture of yeast, yeast food and water, it may be used as yeast to ferment bread, etc.

Base
(a) The bottom portion (of bread, cake, biscuit, etc.).
(b) A baked gâteau, torte or similar cake ready to be decorated.
(c) A substance which reacts with an acid to form a salt.

Basin–Small bowl or container for mixing small quantities of materials.

Batch–The whole of the mixing or contents of an oven.

Batter–The unbaked mixture of ingredients from which cakes, etc., are baked.

Baume, Antoine–French chemist who invented the hydrometer used for measuring sugar syrups (saccharometer). The degrees on the scale are measured in °B. (Baume).

Bay–The hollow arranged in the mixture of dry ingredients into which the liquid is poured preparatory to mixing into the batter for the required cake, etc.

Beat–To agitate ingredients to form a froth, cream or paste.

Beater–That part of the machine which is responsible for agitating the ingredients to form a froth, cream or paste.

B.F.T.–Abbreviation for Bulk Fermentation Time

Blades–The arms of a dough or cake machine which revolve and mix the ingredients.

Blanch–To subject food to boiling water, e.g. the removal of skins of nuts or fruit, scalding fruit for preservation or freezing, etc.

Blend–Mix or distribute materials together.

Blind–
(a) Baking blind is to bake (e.g. a pastry case) without filling.
(b) Cakes, etc., which do not bake with an attractive break on top are said to be "blind".
(c) Non-development of the holes during the baking of crumpets or pikelets.

Blister–Hollow cavities under the top crust of bread or pastry generally caused by baking in too hot an oven.

Bloom
(a) The bright tints of colour produced by caramelization on the crust of baked goods.
(b) The sparkle or lustre seen in flour when viewed in a good light.
(c) Sugar bloom on chocolate shows as a white film on the surface.
(d) Fat bloom on chocolate appears as streaks on the surface.

Blow Degree–The stage at which the temperature of 113°C (235°F) has been reached in a boiling sugar solution.

Board
(a) The table on which doughs and batters are manipulated.
(b) Trays to hold bread and flour confectionery.
(c) Discs or drums, covered in silver or gold paper on which cakes are placed prior to decorating.

Body–Firmness and response to pressure, e.g. of bread crumb, fat, etc.

Border–The decorated edge of cakes.

Bound–Lacking in volume, resulting from a stiff mixing.

Bouquet
(a) Aroma.
(b) Gum paste ornament containing an arrangement of artificial flowers and used for the top decoration of a Wedding cake.

Bowl–Container used for mixing, beating or whisking. If made especially for a machine it is called a machine bowl.

Brake–Mechanical rollers used for rolling or pinning out pastry and biscuit doughs.

Break–The portion of the crust formed during oven spring. It may be on one or two sides of the loaf.

Brioche– A sweet bread or yeast cake made into a fancy shape.
Bun– A small cake usually chemically aerated or fermented.
Bun Wash– A liquid brushed onto fermented buns immediately they are removed from the oven to give them a glaze.
Cake– A baked mixture of edible materials such as flour, fat, sugar, eggs, milk, baking powder, fruit, nuts, etc.
Cake Hoop– Metal ring in which a cake batter is baked.
Cake Tins– Metal shapes in which a cake batter is baked.
Cake Machine– A machine for mixing the ingredients of a cake.
Calorie– The amount of heat required to raise the temperature of 1 gram of water through 1 degree celsius. The kilo calorie is 1000 times this quantity and is used to calculate the energy value of foodstuffs.
Candied– Preserved by a thick layer of sugar crystals, e.g. candied peel.
Caramel
(a) The stage at which sugar takes on an amber colour when heated beyond its melting point.
(b) The stage reached when a boiling sugar solution is heated to beyond 155°C (312°F), and takes on a brown colour.
Caramel Colour– See "blackjack".
Caramel Cutters– These are used to cut a slab of caramel or similar confection into small pieces for chocolate centres. It consists of a metal shaft on which circular cutting discs are assembled with appropriate sized spacers between.
Caramelize– The browning of sugar during the baking of a cake which causes crust colour.
Carbonate of Ammonia– See "Vol".
Carbon Dioxide– The gas produced by fermentation of yeast and chemical aerating agents.
Carbonize– Heating a substance causing chemical changes in it.
Catalyst– A substance which aids a chemical change without itself undergoing any change.
Cell
(a) An enclosed cavity. Can refer to gas or air enclosed in a film of protein material to form cells in bread and cake crumb, meringue, whipped cream, etc.
(b) Structure of all living organisms.
Cellular– Consisting of cells.
Celsius– Temperature scale at which the freezing point of water is 0°, and the boiling point 100° (Formerly the centigrade scale).
Centrepiece– Ornament usually made out of edible materials and used to decorate a large cake, etc.
Centres– Moulded almond paste, jelly, fondant, caramel, nougat, etc., ready for dipping into chocolate, etc., for the making of chocolates and sweets.
Chaffing– Handing up dough to form balls.
Checking– Term used to describe the hair-like cracks which appear on the surface of biscuits.
Chlorination– Chemical treatment of a substance with chlorine, e.g. flour.
Chocolate– Refers only to chocolate couverture – plain or milk.
Chocolate Vermicelli– Granules of chocolate used for decoration.
Chorleywood Bread Process (CBP) – A dough-making process developed by the Flour Milling & Baking Research Association in which the development of the dough is achieved by mechanical energy and certain additives, ascorbic and Vitamin C.
Clarification– Free from impurities. Examples:
(a) Jelly is made more transparent by boiling with egg whites.

(b) Curds in butter are removed by pouring off the oil after heating.

Clearing – Final mixing of ingredients to form a clear dough or batter free from lumps.

Close (Texture) – Fineness of the cellular structure in a texture, not necessarily with lack of volume.

Coagulate – Solidification or partial solidification – usually of protein materials.

Coalesce – Come together to form one, e.g. butter fat in whipped cream.

Coarse (Texture) – Refers to large and rough cells in the texture which are usually irregular but not always so.

Coat – To cover (with icing, paste, etc.).

Cocoa Butter – Fat obtained from cocoa nibs.

Coffin – A black steel box fitted with a lid and in which cakes, choux pastry, etc., may be baked in humid conditions.

Cohesion – Tendency for bodies to remain united, the force holding a solid or liquid together.

Collapse (of cakes) – Breakdown or shrinkage. In cakes resulting in sunken tops.

Comb Scraper – A scraper with a serrated edge, used to pattern the surface of buttercream, icings, etc.

Compote – Sieved fruit, usually cooked in syrup.

Compound Fat – 100% white fat made from hydrogenated lipid oils.

Compress – Squeeze together.

Concave – Curved like the inner surface of a sphere.

Concentrate – Increase the strength of a liquid by reducing its volume.

Concoction – Made up from a number of ingredients.

Condensation – The change of a vapour into a liquid.

Condense – Reduce by evaporation.

Conduction – Method by which heat is transferred through matter i.e. solids.

Cones – Coarsely ground maize or rice used as a dusting medium to prevent soft doughs from sticking during manipulation and proving.

Conserve – Preserve or jam.

Consistency – Firmness and solidity. The criterion by which the proper water absorbtion of flour made into a dough may be judged.

Constituent – Individual component in a recipe.

Contaminate – Pollute or infect.

Contraction – Shrinkage. Reduction in size.

Convection – Method by which heat is transferred in liquids and gases.

Convex – Curved like the outside of a sphere.

Cooler – A device through which bread and cakes, etc., may be rapidly cooled for wrapping.

Coppeaux – Shavings. Can be made from biscuit mixings or chocolate.

Core – The centre or heart.

Cores

(a) Hard pieces or lumps found in the crumb of bread or cakes.

(b) Heavy unbaked portions of the mixing (seams).

Correlate – Connected or mutually related.

Corrosion – Deterioration by chemical action, i.e. rusting of iron.

Cottage Pans – Round bread pans in which cottage or coburg loaves are baked. They can also be used for baking gâteaux bases.

Crack

(a) Break which appears on the surface of cakes and biscuits, etc.

(b) Name used to describe the physical state of a sugar solution when heated above 138°C (280°F).

Cream

(a) Beating two or more ingredients together, to form a slight and fluffy mixture.
(b) Adding cream to a baked cake as a filling or decoration.
(c) Legally the word "cream" must only be used to describe dairy cream.

Cream of Tartar – Potassium Hydrogen Tartrate – one of the acids suitable for use in baking powder.

Cream Powders – Various acids or mixture of acids used in baking powder.

Crimped – Wrinkled or crinkled (to form a decorative edge on pastry, etc.)

Croquant – Another name for nougat. When put through granite rollers, praline paste is made.

Crisp – Brittle or short.

Crown

(a) The top of the oven. So called because the original brick built ovens had an arched top.
(b) System of grading for sultanas 1 to 6.

Crumb – That portion of the loaf other than the crust. The cut surface texture is also usually referred to as the crumb.

Crumbly – Easily rubbed into crumbs. In bread this is a sign of lack of stability in the texture.

Crust

(a) The hard outer portion of bread or cake.
(b) Pastry covering in a pie: Short crust – short pastry.
(c) Top crust is that portion above the pan and break.

Crusty Bread – Loaves baked with a hard brittle crust.

Crystalline – Hard and granular, formed from crystals.

Crystallization – Crystal formation.

Cup Cakes – Small cakes baked in paper cases or cups.

Curd – The protein of milk coagulated by acid or rennet.

Curdle

(a) Formation of curds.
(b) Separation of the ingredients in the emulsion of the cake batter.

Cut Out – To cut out pieces from a mass with a knife or cutter, e.g. marzipan and chocolate.

Cutter – Implement used for cutting shapes out of pastry, etc.

Cutting Back – Kneading the dough part of the way through its bulk fermentation. (Knocking back.)

Cutting the Grain – Converting some of the sugar in a boiling sugar solution to invent sugar by the addition of acid or confectioner's glucose.

Damper

(a) Device used to control the draught of air in a furnace (for heating a baker's oven).
(b) Steam damper – device used to control the amount of steam retained in the oven chamber.

Decorate

(a) To embellish cakes either before or after baking with edible colourful materials such as cherries, nuts, angelica, etc.
(b) Making patterns with icing or buttercream, etc., onto baked cakes, e.g. gâteaux, wedding cakes, etc.

Deep Freeze – Special refrigerator which can rapidly reduce the temperature of both baked and unbaked commodities to below freezing point. Usual deep freeze temperatures are $-23°C$ ($-10°F$).

Dehydrated – Deprived of its water content, e.g. dehydrated egg, etc.

Denatured – Change of the essential qualities, e.g. Gluten which has lost its elasticity.

Deposit – Putting cake batter, etc., into hoops, tins or onto trays. A machine which does this automatically is called a depositor.
Development – State of advancement, i.e. Gluten is developed during the mixing of flour with water.
Divider – Machine which automatically divides off portions of dough of equal size.
Docker – A utensil consisting of a number of spikes and used to puncture the surface of a sheet or shape of dough or paste.
Dormant – Lying inactive and awaiting suitable conditions to become active.
Double Saucepan – The principle of the bain-marie applied to a saucepan.
Dough – Mixture of flour and water, which may or may not be fermented.
D'Oyley – A fancy lace-like mat used for presenting foodstuffs. It may be made of paper, plastic or fabric.
Dragees – Small ball-like sweets coloured either gold or silver, used in the decoration of cakes.
Draw – Remove baked goods from the oven.
Dredger – Utensil used to sprinkle flour, icing sugar, etc.
Drumming – Stretching greaseproof or silicone paper across a cake hoop and securing by twisting the edge over the rolled edge of the hoop. This prevents the leakage of soft batters and adds more protection from heat to the bottom outside edge of the cake.
Dust – To sprinkle flour, etc., onto a workbench to prevent doughs sticking, or onto goods either before or after baking.
E.E.C. – Abbreviation for European Economic Community.
Egg Wash – Egg usually diluted with water, used to wash dough and pastry to produce a glaze when baked.
Elasticity – The property of a material by which it tends to recover its original shape after deformation, e.g. gluten in a well manipulated dough.
Emulsifier – A machine which breaks down an oil and water mixture to very small particles, resulting in a stable emulsion.
Emulsifying Agent – Substance which gives stability to an emulsion, e.g. lecithin.
Emulsion – Mechanical mixture of fluids which do not naturally mix, e.g. oil and water.
Enrichment – The addition of enriching agents such as fat, eggs, etc.
Enrobe – Coating of cakes, etc., with icing. The machine which does this automatically is called an enrober.
Enzymes – Organic catalysts responsible for many of the chemical changes which occur in food.
Fahrenheit – Temperature scale at which the freezing point of water is 32° and the boiling point is 212°.
Fancies – Small decorated cakes.
Feather Degree – The stage at which the temperature of 115°C (240°F) has been reached in a boiling sugar solution.
Ferment – Mixture of flour, sugar, yeast, yeast food and water which is allowed to ferment prior to it being made into a dough by the addition of flour and other ingredients. This is recommended for rich fermented doughs.
Fermentation – The action of yeast in a dough in which the sugar is changed into carbon dioxide gas and ethyl alcohol.
Final Proof – The stage between the final moulding and the baking of a dough.
Firing – Applying heat to an oven.
Flaked – Slices cut very thinly.
Flan – Open pastry or sponge case with a suitable filling.
Flash – Place into a very hot oven to colour, e.g. meringue piped on as decoration, rout biscuits, etc.

Flash Heat – Fierce heat which burns the surface of goods in an oven, before they are properly baked.
Flat Sheet – Ordinary baking sheet.
Fluted – Having vertical grooves.
Fondant – A special icing made from boiling sugar.
Fortified – The addition of nutrients.
French Knife – A sharp knife in which the end tapers to a point.
Full Proof – Maximum expansion of a dough before collapse.
Fungicide – Substance which destroys fungus.
Ganache – A mixture of boiled dairy cream and chocolate couverture.
Garnish – Embellish, e.g. parsley on savoury goods.
Gâteau – French term for a decorated cake.
Gel – A semi-solid colloidal solution.
Gelatine – A transparent jelly sold in powder or sheet form and used for making jellies, creams, etc.
Gelatinization – Formation of a gel.
Genoese – Good quality plain cake baked in a sheet and cut up for making fancies, layer cakes, gâteaux, etc.
Glaze – To cover with something which imparts a glossy surface.
Gloss – The shiny reflective surface on chocolate, fondant, etc.
Glucono Delta Lactone (G.D.L.) – An agent substituting for the acid used in baking powder.
Glucose (*Confectioner's*) – The viscous syrup (inc. dextrose) used in sugar boiling to prevent premature crystallization.
Gluten – Hydrated wheat protein which can be washed out of a dough made from flour and water.
Glycerine – A colourless sweet-tasting syrup used in cakes for its hygroscopic properties which prolongs their shelf life.
Grain – In bread this refers to the size, shape, and arrangement of the cells comprising the cut surface of the crumb.
Grained – Crystallization of a boiled sugar solution.
Green Dough – Unripe dough, not having received sufficient fermentation.
Grease – Brush or spread fat into baking tins or trays.
Gum Paste – A paste made from icing sugar, gum tragacanth, cornflour and water.
Gums (*Also see Stabilizers*) – In the food industry there are a whole range of gums which are used to improve food quality by imparting desirable viscosity, texture, moisture-weepage control, gel formation, opacity or clearness, and freeze/thaw stability. The following list is by no means complete but are in present use:

Agar, Alginate, Arabic, Carrageen, Gelatine, Ghatti, Guar, Karaya, Locust Bean, Pectin, Tragacanth, Xanthan.

Hair Sieve – A sieve made of horsehair used for sieving soft fruits.
Handing Up – Preliminary moulding of dough pieces prior to recovery and final moulding.
Hard Ball – The stage at which the temperature of 121°C (250°F) has been reached, in a boiling sugar solution.
Hard Crack – The stage at which the temperature of 270–280°C (132–138°F) has been reached in a boiling sugar solution.
Head (*On a Ferment*) – The foam resulting from the fermentation of a liquid or semi-liquid medium.
Harp – A utensil used to trim genoese.
High Ratio Cake – Refers to cakes containing high ratios of sugar and moisture to flour.

Fat. Edible 100% shortenings containing emulsifying agents suitable for use for high ratio cakes.

Flour. Specially finely milled high grade chlorinated flour suitable for use for high ratio cakes.

High Protein, High Ratio Flours – High ratio flour milled from stronger wheat suitable for high ratio cakes which contain fruit.

High Speed Mixers – These mix ingredients at about 3 times the speed of conventional mixers.

Homogenize – To break up the fat particles of milk into smaller particles.

Homogenizer – Machine which homogenizes milk.

Hoop – A circular band of metal in which cakes are baked.

Hopper – A funnel into which flour, etc., is passed from an upper level, usually through a sifter into receptacles at a lower level.

Hot Air Ovens – Ovens heated by recycled hot air from a furnace through a series of ducts.

Hotplate – A flat metal surface which may be heated and on which goods such as scones, crumpets and muffins are baked.

Humidity – The moisture in the atmosphere.

Hydration – Combination of substances with water.

Hydrogenation – Process of adding hydrogen gas (in the presence of a catalyst) to convert liquid oils into solid fats.

Hygrometer – Instrument for determining the relative humidity (degree of saturation of moisture in air).

Hygroscopic – Absorbtion of moisture.

Icing – A coating and/or decorating medium, e.g. royal icing, water icing, fondant, etc.

Icing Sugar – Finely powdered sugar from which icings may be made.

Immature Dough – A dough having insufficient fermentation.

Immiscible – Will not mix.

Impermeable – That which will not allow the passage of liquids.

Impregnated – Filled or saturated.

Improvers – Substances which are added to the basic bread or flour confectionery recipe to improve its quality or shelf life.

Incision – A cut or gash.

Incorporate – To combine, e.g. ingredients in a recipe.

Infection – Transfer of a disease.

Infestation – Contaminated with vermin of pests, etc.

Infusion – Extract obtained by steeping substances in a liquid solvent.

Inhibit – Prohibit, forbid, restrain, etc.

Initiate – Originate, begin.

Injection – A liquid forced under pressure into something.

Insipid – Devoid of flavour, uninteresting.

Insoluble – Cannot be dissolved.

Instant – Term applied to many dried foods which can be rapidly reconstituted, i.e. instant coffee.

Inversion – The change from sucrose sugar to invert sugar (glucose/fructose) which occurs during the boiling of a sugar solution or by the action of the enzyme invertase.

Invertase – Name of the enzyme responsible for the above change of sucrose into Invert Sugar.

Invert sugar – A mixture of glucose (dextrose) and fructose (laevulose).

Jellying Agents – Substances which cause a liquid to set, e.g. Agar, gelatine, pectin, gum tragacanth, etc.

Jigger Wheel – Small serrated wheel made of metal or wood used to cut paste into strips simultaneously leaving a decorative edge.

Joule – Unit of work done or heat generated by a current of one ampere, acting for one second against a resistance of one ohm. The E.E.C. have now adopted this unit instead of calories for measuring energy values in foodstuffs
1 Joule = 4·12 kilocalories.

Kind – A term used in bakery to describe the satisfactory feel of a dough or appearance of a cake, e.g. a "kind" break or crack.

Kirsch – A liqueur distilled from cherries and used for flavouring purposes, especially for torten.

Knead – Work a piece of dough by vigorous manipulation.

Kneader – Machine able to mix a dough.

Knock Back – Expelling the gas from fermented dough usually about $\frac{2}{3}$rds through its bulk fermentation, by kneading or folding, etc.

Lagging – Insulating material wrapped round pipes etc. to prevent loss of heat (or cold).

Lamination – Formation of layers, e.g. puff pastry.

Latent Heat – The heat required or evolved to change the state of a substance, e.g. solid to a liquid or liquid to gas (or vice versa) without any change in temperature.

Lecithin – A stabilizer used in emulsions. Present in soya bean and egg yolk.

Lift – Refers to the expansion of the layers in puff pastry.

Lintner Value – Measure of the capacity of a malt to convert starch into sugar, i.e. diastatic value.

Liquefy – Change into the liquid state.

Liqueurs – Strong sweet alchoholic spirits. Used in flour confectionery for flavouring purposes, e.g. Kirsch.

Liquidizer – Machine or an attachment to a cake machine, which will pulverize foodstuff to a puree or liquid.

Liquor – Technical term used in bakeries to describe water or solutions of sugar, salt etc.

Live Cells – Living yeast cells.

Lively – Refers to a dough or ferment in which the fermentation is vigorous.

Loaf – A piece of dough moulded into the required shape, proved and baked. In Britain some sizes are dictated by Law.

Long Patent – Second grade white flour.

"M" Fault – Term used to describe the faults which cause the top of a cake to collapse.

Macerate – Steep, mix or mash. Dissolve in a liquid solvent.

Machine Doughs – Doughs mixed by machine.

Maidenhair Fern – Dried and pressed fern used in decoration especially with marzipan roses.

Major Factor – Method used to calculate the water temperature for doughs (*see* page 49).

Maltase – The enzyme responsible for converting maltose sugar into glucose sugar.

Maltose Figure – The percentage of sugars in flour, calculated as maltose, plus that which may be produced from the damaged starch cells during fermentation and baking.

Manipulation – Working (a dough) by hand or machine.

Maraschino – A liqueur made from a sour black cherry grown in Dalmatia.

Marbled Icing – Decorative effect produced by the skilful manipulation of different coloured icings piped onto an iced surface.

Marmalade – A preserve made from oranges, lemons, grapefruit, sugar and their rinds.

Marzipan – A paste made from almonds and sugar. Legally if sold as such, it should contain not less than 23·5% of dry almond substance and no other nut ingredient. Not less than 75% of the remainder shall be solid carbohydrate sweetening matter.

Masking – Covering with icing, etc.
Mat Surface – Dull, the converse of glossy.
Mature – Ripe – fully developed.
Maw Seeds – The seeds from a variety of poppy used for the decoration and flavour of certain types of bread and rolls. Can be purchased as "blue" or "white".
Meal Worm – An insect which feeds on cereal products.
Mean – (Temperature, etc.) Average.
Measuring Tank – Tank used for measuring the correct amount of water for a dough.
Mediterranean Flour Moth – Pest, the larvae of which form large quantities of webbing which becomes a nuisance in flour.
Melangeur – A mill consisting of two granite rollers which rotate in a cylindrical trough. Materials such as cocoa nibs are reduced to a paste in such machines.
Melting Point – The temperature at which a solid substance like fat melts and becomes liquid.
Mesh – The small orifices through which materials are allowed to pass in a sieve.
Meter – An apparatus by which measurements are made.
Microbe – Any micro-organism.
Micro-organism – Any organsim which cannot be seen without the aid of a microscope.
Mill – Grinding machine. A flour mill consists of other ancillary machinery which reduce wheat to various types of flour.
Miller – Person in charge of the flour mill.
Milling – Grinding a hard substance into a fine powder.
Mincemeat – Mixture of various finely chopped dried fruits, apple, suet, sugar and spices usually with rum or brandy, used as a filling for pies and tarts.
Mineral Matter – Minerals found naturally in foodstuffs as determined by their ash content.
Miscible – Able to form a perfect mixture.
Mites – Small insects which thrive in flour.
Mixer – A machine which mechanically mixes the ingredients of a recipe.
Mixing Times – Refers to the time of mixing various doughs, etc.
Mould –
(1) Form into a shape, e.g. dough.
(2) Hollow shape into which a plastic substance, etc. can be modelled.
(3) Micro-organism which thrives in a damp environment.
Moulder – Machine used for shaping pieces of dough.
Mouldy – Contaminated with mould.
Mucilage – A colloidal secretion found in plants, e.g. mucilage of gum tragacanth.
Musty – A taint which develops in flour etc. if stored in unsuitable conditions.
Nibs – Small fragments of materials, e.g. sugar or almond nibs.
Nippers – Pincers used to impart a decorative pattern on various pastes.
Noisette – Name given to pastes made from or containing roasted nuts.
Nonpareils – Fine short threads of coloured sugar or chocolate used for decoration.
Non-stick Baking Tins – Tins which are coated with silicone resins and cured by heat.
No-time Doughs – Doughs which have no bulk fermentation.
Nougat –
(1) A mixture of boiled sugar and almonds/hazelnuts. Can be ground to form a paste (praline paste).
(2) Montelimart, confection made from boiled sugar, honey and egg whites with added fruits, angelica and nuts.
Noyeau – An almond flavoured alcoholic liqueur.
Nutrient – A substance which provides nourishment, i.e., carbohydrates, lipids, proteins, minerals, and vitamins.

Old Dough–Dough which is over fermented (over-ripe).

Orange Flower Water–A distillate from the flower of the orange tree used as a flavouring agent especially in almond paste.

Oven–Equipment in which goods can be baked.

Crown–Arched roof of the baking chamber.

Peel–Flat shovel made of wood or metal fitted to a long handle, used to remove goods from the oven.

Stock–Front of a brick built oven.

Tube–Steam tube used to heat this type of oven.

Oven Spring–The expansion in size gained during the baking of a loaf of bread.

Over-Fermented–Fermented beyond the optimum period from which good results can be obtained.

Over-Proof–Proved beyond the optimum time from which good results can be obtained.

Overrun–Increase of volume which occurs when fresh or imitation cream is whipped.

Oxidation–The addition of oxygen or removal of hydrogen from a substance.

Palette Knife–Flexible thin, blunt, round-ended knife used for spreading purposes. A trowel or drop palette knife has the blade lower than the handle to facilitate spreading mixtures in a tin with raised edges.

Pans–Baking tins.

Pan Cottage–Cottage loaf baked in a tin.

Panary Fermentation–Fermentation of a bread dough.

Parboil–Partly boil.

Pasteurization–Process by which foods are heat treated to arrest the development of bacteria.

Pastillage–Paste made from icing sugar and gelatine mucilage.

Pastries–Products made by the pastrycook or confectioner.

Pasty–A small pastry containing meat and vegetables.

Pastry Bake–Machine used to roll out pastry.

Patents–Term used for different grades of white flour, e.g. top patent, bottom patent, etc.

Patty–Small pastries containing a savoury filling.

Peaked Tops–A fault in cakes which manifests itself as a peak with cracks on top.

Pearl Degree–The temperatures of 110°C (230°F) in a boiling sugar solution.

Pectin–The naturally occurring jellying agent found in certain fruits.

Peel–

(1) See oven peel.

(2) The rind of citrus fruits.

Persipan–A substitute marzipan or almond paste made from apricot or peach kernals.

Pervious–Penetrable, open or perforated.

Petits Fours–Very small fancy cakes which may be either iced (glâcé) or left dry (sec).

Pie–Meat or fruit baked in a dish with a pastry lid or in a pastry case.

Pile–Term used to describe the build up of the crumb of bread.

Pin–Rolling pin.

Pincers–Fancy-shaped gripping tool used for putting a decorative pattern onto various types of pastes.

Pinched–Decoration applied to the edge of shortbread by the fingers.

Pinning–Rolling out with a rolling pin.

Piping–Forcing of an icing through a tube in a bag. By the use of special tubes, and skilful manipulation of the bag, artisitic decorative effects can be made.

Plait–Various number of ropes of dough woven together.

Planetary Action– Method by which mixing is achieved by a normal vertical cake machine.
Plant Bakery– A bakery in which all the operations involved in the making of bakery goods is done by machine.
Plaque– Ornamental tablet.
Plaster Moulds– Moulds made in plaster of paris (calcium sulphate) from which sugar or almond paste figures, etc., may be cast.
Poach– Simmer-heat just below boiling point until cooked.
Pound Cakes– Originally referred to cakes which were made from 1 lb of each of butter eggs, sugar and flour.
P.O.E.M.S.– Polyoxyethylene monostearate – an anti-staling or crumb softening agent for bread.
Praline– Nougat (or croquant) which has been made into a paste by passing it through rollers.
Precipitate– Insoluble matter thrown out of solution.
Proof– Aeration imparted by the action of yeast in fermentation of dough immediately prior to goods being baked.
Proteolysis– Action of proteolytic enzymes.
Proteolytic Enzymes– Enzymes which split the large protein molecules into simpler, smaller amino-acid molecules.
Prover– Cabinet into which steam and heat can be applied and in which fermented goods are proved prior to baking.
Pulled Sugar– Sugar syrup which has been boiled with acid or glucose to 156°C (312°F) and when cool enough to handle is pulled until it obtains a satin-like sheen. It may be manipulated to form flowers, fruits and figures which are used for decoration.
Pumpable shortenings– Shortenings which are manufactured to a consistency suitable to be pumped into continuous cake making machinery.
Puree– A smooth sieved pulp
Rack Oven– A type of oven in which a rack is made to revolve in a vertical position inside a heated chamber.
Radiation– Transfer of heat by heat waves or rays.
Raised Pies– Pies in which the pastry is "raised" either by hand or machine to form the hollow case into which the filling is baked.
Reaction– Chemical change which takes place between substances.
Ream– 480 (or 500) sheets of paper.
Recipe– The exact formula giving the quantities of ingredients to use and method for making a particular type of bread or confectionery.
Recovery Time– Time which is allowed for dough to relax between various manipulative operations.
Reduction– Term used by scientist to explain the removal of oxygen or other electro-negative atom or group from a compound.
Reel Ovens– A type of travelling oven in which each shelf is made to revolve in a horizontal position inside a heated chamber.
Refiner– Machine used for purifying.
Refraction– The bending of a path of light waves travelling obliquely from one medium to another, occurring at the surface separating the two media.
Refrigerants– Substances which induce refrigeration.
Relative Humidity– In simple terms this means the percentage of moisture present in air, e.g. 100% RH = Air saturated with moisture. 50% RH = Air $\frac{1}{2}$ saturated with moisture.
Rennet– A substance which contains the enzyme rennin, used for coagulating milk to make cheese and curd.

Retarder–Cabinet which is kept at a temperature of 2°–4°C (38°–40°F) and RH of 80% and into which unbaked fermented goods may be held (retarded) up to 24 hours prior to baking.

Rich Cakes–Those of the highest quality, e.g. richly fruited cakes.

Rinds–The peel of fruit or vegetables.

Ripe Dough–Dough which has reached the optimum degree of fermentation to produce good results.

Ripening (of Gluten)–The softening and conditioning which takes place during fermentation.

Rock Sugar–A rock-like material made by stirring royal icing into a boiled sugar syrup and used to decorate cakes, especially Christmas cakes.

Rochelle Salt–Potassium sodium tartrate which is the residual salt from the chemical action of sodium bicarbonate and cream of tartar baking powder.

Rope–A disease of bread which produces a sticky crumb which can be pulled away in strands.

Rose Water–Product from the steam distillation of rose petals and used for flavouring purposes.

Rotary Oven–A type of travelling oven containing swinging shelves.

Rotten Dough–Dough which is so over-ripe that the gluten becomes short in character.

Roux–A cooked mixture of flat and flour used for making sauces.

Royal Icing–Mixture of icing sugar and egg whites beaten together and used for decorative purposes.

Run-Outs–Dried shapes made from royal icing to a pattern and used in cake decoration.

Runny Dough–Dough lacking in stability.

Saccharometer–A hydrometer calibrated in degrees. Beaume (or Brix) used to determine the relative density of sugar solutions.

Sack–Before metrication, the standard unit by which flour was sold in the United Kingdom. The weight of a sack was 280 pounds (127 kg).

Saffron–The dried stigmas of the saffron crocus which when made into an infusion is used for colouring and flavouring purposes.

Saline–Salty.

Salt–The common salt used in recipes for flavour is sodium chloride. In chemistry the name "salt" is also applied to the substance produced when an acid reacts with a base. (acid + base = water + salt).

Sandarac–A gum used to make edible varnish.

Sand Cake–Cake in which the flour used is predominately cornflour.

Sandwich Loaf–Square-shaped loaf baked in a tin suitable for cutting for sandwiches.

Sandwich Plates–Round tins in which sponge sandwiches are baked.

Saturated Solutions–Solutions in which the maximum amount of solute remains in solution without being precipitated at normal room temperature.

Savoy Bag–Cone-shaped bag made of cloth, nylon, plastic, etc., fitted with a replaceable nozzle through which mixings may be piped out.

Savoy Tubes–Nozzles of various sizes and patterns which may be fitted into the savoy bag.

Scales–A weighing instrument.

Scaling Off–Weighing off pieces of dough or portions of mixings on a pair of scales.

Scoop–Utensil like a small shovel used for moving small quantities of raw materials, e.g. from the store bin to the scales.

Scrap–Scrapings and rubbings from a dough or mixing.

Scraper–Utensil used to scrape the sides of a bowl free of mixing. Usually made of plastic or celluloid.

Chisel. Narrow steel blade with a wooden handle used for scraping mixings like nougat, etc., from a slab on which it is poured.
Comb. Scraper which has a serrated edge used to impart a decorative pattern when used to spread decorating mediums such as buttercream, chocolate, etc.
English. Tool which consists of a semi-circular steel blade set at and angle of 90° from the handle which is made of wood. Used for marzipaning cakes.
Scotch. Flat wide steel blade with a wooden handle, used for scraping baking tins and dividing scones, rounds etc.

Scroll – Decorative shape used to embellish cakes.
Scuffle – Long pole on which a piece of cloth or sacking is attached. Used for cleaning the sole of internally fired ovens.
Scum – Froth or extraneous matter which rises to the surface of liquids when boiled.
Seams – The join between two edges of dough, etc., especially in the moulding of a shape.
Season – Dulling the shiny surface of new tins and pans by putting them in a hot oven for a few hours, i.e. not less than 226°C (440°F). This is to facilitate the penetration of heat during baking.
Seasoning – The adding of salt, pepper, spices, etc., to savoury foods.
Sediment – Insoluble substances that separate and sink to the bottom.
Set – Batch of bread (or cakes).
Setting – Filling an oven with bread or cakes.
Setter – Strip of wood to hold a row of loaves during final proof, prior to baking.
Sheen – Sparkle which can be seen on the cut surface of a well fermented loaf of bread.
Sheet Tin – Metal trays on which goods are baked.
Shell Top – The top crust of baked goods separating during baking to form a kind of cap.
Shortpastry – A friable pastry made from flour and fat, etc.
Sieve – Utensil with a mesh of wire, nylon, etc., through which dry materials may be passed to remove large particles or extraneous matter.
Skinning – The hard crust which forms on dough when left uncovered.
Slab Cake – Plain or fruited cake baked in a large square or rectanglar tin or frame.
Slack Dough – A soft dough usually containing extra water.
Smoke Point – Temperature at which frying oils emit a bluish coloured smoke.
Snow – Term used to describe the white foam which results when egg whites are whipped.
Sodium Bicarbonate – The ingredient of baking powder which when reacted with an acid liberates carbon dioxide gas which aerates the goods.
Soft Ball Degree – The stage at which the temperature of a boiling sugar syrup reaches 115–118°C (290 – 245°F).
Soft Crack Degree – The stage at which the temperature of a boiling sugar syrup reaches 132°C (270°F).
Soft Flour – A flour containing a weak or low percentage of gluten.
Sole – The floor of an oven.
Solid Heat – Heat which has been absorbed by the fabric of the oven so that goods can be baked without too great a drop in the baking temperature.
Sour Dough – Refers to dough in which considerable quantities of acid has been produced.
Spatula – Flat wooden implement used for stirring or beating small quantities of mixtures.
Specific Heat – The number of calories required to raise 1 gram of a substance through 1 degree celsius. Specific heat of water = 1. Specific heat of flour = 0·45 calories per gram per °C.

Specific Heat Capacity – The quantity of heat required to raise the temperature of unit mass of a substance by one degree. Expressed in joules per kilogram per kelvin. SI units.

Splash – Spraying a liquid over goods, e.g. water or liqueur (for torten).

Split –
(1) Divide into two equal parts.
(2) Refers to a special cut e.g. split almonds.

Spindle Moulder – Machine which moulds tin bread by means of a revolving spindle.

Sponge –
(1) Thin batter made from flour, water, salt and yeast which is allowed to ferment and then made into a dough with more flour etc.
(2) Mixture of egg and sugar beaten to a stiff foam and flour added.

Spun Sugar – Sugar syrup boiled to the hard crack degree and spun into thin threads.

Stabilizer (*Also see Gums*) – Colloidal substances used to stabilize emulsions, i.e. prevent separation of oil and water.

Stability (*of Dough*) – Ability of a dough to retain its shape during fermentation of the moulded dough piece.

Star Tubes – Piping tubes in which the hole is cut in such a way as to leave currugations in the mixture as it is forced through.

Steep – Soak in water.

Stencil – Thin material into which is cut a pattern which may be reproduced in icing etc. on cakes.

Sticky Doughs – Sticky to touch resulting from under-ripe doughs, lack of salt etc.

Stock –
(1) All the materials used, usually kept in a storeroom.
(2) Liquor produced from the stewing of bones etc. used as the basis for making soups, sauces, etc.
(3) Syrup – Simple syrup made from sugar and water used for reducing fondant, etc.

Straight Dough – Method of making a dough in which all the ingredients are blended at one operation.

Straight Run Flour – One grade of flour which consists of all the white flour obtained from the wheat by the miller, usually 72–75% extraction.

Streaky Texture – Layers of different coloured or more dense texture which show up as streaks.

Strong Flour – A flour containing a strong or high percentage of gluten.

Sugar Paste – A paste made from sugar and a jellying agent used for decorative purposes.

Super-Saturated Solution – Solutions which carry more sugar than ordinary saturated ones.

Taint – Infect or stain with anything noxious.

Tart – A pastry case baked with a suitable filling, e.g. jam, custard, etc.

Tempering – Process by which chocolate couverture is conditioned prior to its use.

Tempering Tank – Tank in which the water at the required temperature is measured prior to mixing with the other ingredients to make dough.

Template (*Templet*) – Pattern used as a guide for work to be done of a decorative nature, e.g. decoration of cakes in royal icing.

Texture – Structure. In bread and confectionery it relates to the cut surface of the loaf or cake.

Textured Vegetable Protein (*T.V.P.*) – Made from the protein of soya, this product can be used to replace a percentage of meat in savoury products. Several different sizes are available. It is made in a coloured or neutral type and in several different savoury flavours.

Thread–Refers to way desiccated coconut can be cut i.e. in long threads.

Thread Degree–The stage at which the temperature of a boiling sugar syrup reaches 110°–113°C (230–233°F).

Tight–Dough or mixing containing insufficient liquor.

Top Patent Flour–Highest grade of white flour obtainable from the miller.

Tough–Tenacious, strong or inextensible, e.g. gluten from a strong flour is said to be tough.

Toughen–To make tough, e.g. by manipulation.

Trade Descriptions Act–An Act of Parliament put on the Statute Book in the United Kingdom in 1968. It is intended to protect the consumer against the use of false descriptions with regard to the sale of goods.

Trough–Open wooden vessel in which bread doughs were originally made by hand, but is now often used to contain dough during its bulk fermentation time.

Turntable–A piece of equipment used to rotate a cake to facilitate coating and piping designs in icing.

Twists–Bread formed from two or more ropes of bread.

Umbrella Moulder–A type of bread moulder.

Under-Proving–Insufficient proof allowed.

Under-Ripeness–Insufficient fermentation allowed.

Utensil–Tool or instrument used in the preparation of foods.

Viscosity–The resistance shown by a fluid to flow. Dough can be regarded as a semi-fluid in this respect.

Vol–Bakery term for an aerating agent comprising ammonium carbonate and ammonium hydrogen carbonate (bicarbonate).

Wafer Paper–Edible sheets of paper made from gelatinized starch and gum, upon which certain goods such as macaroons may be baked.

Wash–Liquid used for glazing or moistening confectionery, e.g. egg wash.

Well–See "Bay".

Whip–Beat with a whisk.

Whisk–Bundle of wires arranged so that they may be handled and used as a beater.

Wires–A tray made of wires and used for draining, cooling etc., of various confectionery goods.

X Fault–The fault in cakemaking which manifests itself as a collapsed top and shrunken sides. Caused by excessive liquid.

Yeast–Living micro-organism which under favourable conditions of warmth, moisture and food will grow and produce carbon dioxide gas and alcohol.

Yield–The number of units which a particular recipe is calculated to give.

Zest–Outer portion of peel which contains the essential oil of the fruit.

Zymase–Group of enzymes in yeast which break down glucose sugar into carbon dioxide gas and ethyl alcohol.

Appendix 1

TEMPERATURE CONVERSIONS

There are two methods by which the degrees in one scale can be converted into another. To change Celsius (Centigrade) into Fahrenheit:

Method 1

Multiply by $\frac{9}{5}$ and add 32.

Method 2

Add 40, multiply by $\frac{9}{5}$ and subtract 40.

To change Fahrenheit into Celsius:

Method 1

Subtract 32 and multiply by $\frac{5}{9}$.

Method 2

Add 40, multiply by $\frac{5}{9}$ and subtract 40.

Examples

(1) Change 25°C into Fahrenheit.

Method 1

$25 \times \frac{9}{5} = 45$.
$45 + 32 = 77$°F (Answer)

Method 2

$25 + 40 = 65$
$65 \times \frac{9}{5} = 117$
$117 - 40 = 77$°F (Answer)

(2) Change 41°F into Celsius

Method 1

$41 - 32 = 9$.
$9 \times \frac{5}{9} = 5$°C (Answer)

Method 2

$41 + 40 = 81$
$81 \times \frac{5}{9} = 45$
$45 - 40 = 5$°C (Answer)

A temperature conversion table appears opposite.

Conversion Tables

These tables will be useful for recipe conversion.

The numbers in heavy type can be either °C or °F. If the heavy type number is °C, the equivalent in °F is on the right. If the heavy type number is °F, the equivalent in °C is on the left.

°C		°F	°C		°F	°C		°F	°C		°F
−40	**−40**	−40	0·5	**33**	91·4	21·0	**70**	158·0	71	**160**	320
−34	**−30**	−22	1·1	**34**	93·2	21·5	**71**	159·8	76	**170**	338
−29	**−20**	− 4	1·6	**35**	95·0	22·2	**72**	161·6	83	**180**	356
−23	**−10**	+14	2.2	**36**	96·8	22·7	**73**	163·4	88	**190**	374
−17·7	**− 0**	+32	2·7	**37**	98·6	23·3	**74**	165·2	93	**200**	392
−17·2	**1**	33·8	3·3	**38**	100·4	23·8	**75**	167·0	99	**210**	410
−16·6	**2**	35·6	3·8	**39**	102·2	24·4	**76**	168·8	100	**212**	413
−16·1	**3**	37·4	4·4	**40**	104·0	25·0	**77**	170·6	104	**220**	428
−15·5	**4**	39·2	4·9	**41**	105·8	25·5	**78**	172·4	110	**230**	446
−15	**5**	41·0	5·5	**42**	107·6	26·2	**79**	174·2	115	**240**	464
−14·4	**6**	42·8	6·0	**43**	109·4	26·8	**80**	176·0	121	**250**	482
−13·9	**7**	44·6	6·6	**44**	111·2	27·3	**81**	177·8	127	**260**	500
−13·3	**8**	46·4	7·1	**45**	113·0	27·7	**82**	179·6	132	**270**	518
−12·7	**9**	48·2	7·7	**46**	114·8	28·2	**83**	181·4	138	**280**	536
−12·2	**10**	50·0	8·2	**47**	116·6	28·8	**84**	183·2	143	**290**	554
−11·6	**11**	51·8	8·8	**48**	118·4	29·3	**85**	185·0	149	**300**	572
−11·1	**12**	53.6	9·3	**49**	120·2	29·9	**86**	186·8	154	**310**	590
−10·5	**13**	55·4	9·9	**50**	122·0	30·4	**87**	188·6	160	**320**	608
−10·0	**14**	57·2	10·4	**51**	123·8	31·0	**88**	190·4	165	**330**	626
−9·4	**15**	59·0	11·1	**52**	125·6	31·5	**89**	192·2	171	**340**	644
−8·8	**16**	60·8	11·5	**53**	127·4	32·1	**90**	194·0	177	**350**	662
−8·3	**17**	62·6	12·1	**54**	129·2	32·6	**91**	195·8	182	**360**	680
−7·7	**18**	64·4	12·6	**55**	131·0	33·3	**92**	197·6	188	**370**	698
−7·2	**19**	66·2	13·2	**56**	132·8	33·8	**93**	199·4	193	**380**	716
−6·6	**20**	68·0	13·7	**57**	134·6	34·4	**94**	201·2	199	**390**	734
−6·1	**21**	69·8	14·3	**58**	136·4	34·9	**95**	203·0	204	**400**	752
−5·5	**22**	71·6	14·8	**59**	138·2	35·5	**96**	204·8	210	**410**	770
−5·0	**23**	73·4	15·6	**60**	140·0	36·1	**97**	206·6	215	**420**	788
−4·4	**24**	75·2	16·1	**61**	141·8	36·6	**98**	208·4	221	**430**	806
−3·9	**25**	77·0	16·6	**62**	143·6	37·1	**99**	210·2	226	**440**	824
−3·3	**26**	78·8	17·1	**63**	145·4	37·7	**100**	212·0	232	**450**	842
−2·8	**27**	80·6	17·7	**64**	147·2	43	**110**	230	238	**460**	860
−2·2	**28**	82·4	18·2	**65**	149·0	49	**120**	248	243	**470**	878
−1·6	**29**	84.2	18·8	**66**	150·8	54	**130**	266	249	**480**	896
−1·1	**30**	86·0	19·3	**67**	152·6	60	**140**	284	254	**490**	914
−0·6	**31**	87·8	19·9	**68**	154·4	65	**150**	302	260	**500**	932
−0	**32**	89·6	20·4	**69**	156·2						

The numbers in heavy type can be either litres or pints. For example, if the heavy type number is 5, 5 litres are equivalent to 8·7990 pints and 5 pints are equivalent to 2·8412 litres.

Litres – Pints

(8 pints = 1 imperial gallon)

litres		pints	litres		pints	litres		pints
0·5682454	**1**	1·7598	19·8886	**35**	61·5930	38·6407	**68**	119·6664
1·1365	**2**	3·5196	20·4568	**36**	63·3528	39·2089	**69**	121·4262
1·7047	**3**	5·2794	21·0251	**37**	65·1126	39·7772	**70**	123·1860
2·2730	**4**	7·0392	21·5933	**38**	66·8724	40·3454	**71**	124·9458
2·8412	**5**	8·7990	22·1616	**39**	68·6322	40·9137	**72**	126·7056
3·4095	**6**	10·5588	22·7298	**40**	70·3920	41·4819	**73**	128·4654
3·9777	**7**	12·3186	23·2981	**41**	72·1518	42·0502	**74**	130·2252
4·5460	**8**	14·0784	23·8663	**42**	73·9116	42·6184	**75**	131·9850
5·1142	**9**	15·8382	24·4346	**43**	75·6714	43·1867	**76**	133·7448
5·6825	**10**	17·5980	25·0028	**44**	77·4312	43·7549	**77**	135·5046
6·2507	**11**	19·3578	25·5710	**45**	79·1910	44·3231	**78**	137·2644
6·8189	**12**	21·1176	26·1393	**46**	80·9508	44·8914	**79**	139·0242
7·3872	**13**	22·8774	26·7075	**47**	82·7106	45·4596	**80**	140·7840
7·9554	**14**	24·6372	27·2758	**48**	84·4704	46·0279	**81**	142·5438
8·5237	**15**	26·3970	27·8440	**49**	86·2302	46·5961	**82**	144·3036
9·0919	**16**	28·1568	28·4123	**50**	87·9900	47·1644	**83**	146·0634
9·6602	**17**	29·9166	28·9805	**51**	89·7498	47·7326	**84**	147·8232
10·2284	**18**	31·6764	29·5488	**52**	91·5096	48·3009	**85**	149·5830
10·7967	**19**	33·4362	30·1170	**53**	93·2694	48·8691	**86**	151·3428
11·3649	**20**	35·1960	30·6853	**54**	95·0292	49·4373	**87**	153·1026
11·9332	**21**	36·9558	31·2535	**55**	96·7890	50·0056	**88**	154·8624
12·5014	**22**	38·7156	31·8217	**56**	98·5488	50·5738	**89**	156·6222
13·0696	**23**	40·4754	32·3900	**57**	100·3086	51·1421	**90**	158·3820
13·6379	**24**	42·2352	32·9582	**58**	102·0684	51·7103	**91**	160·1418
14·2061	**25**	43·9950	33·5265	**59**	103·8282	52·2786	**92**	161·9016
14·7744	**26**	45·7548	34·0947	**60**	105·5880	52·8468	**93**	163·6614
15·3426	**27**	47·5146	34·6630	**61**	107·3478	53·4151	**94**	165·4212
15·9109	**28**	49·2744	35·2312	**62**	109·1076	53·9833	**95**	167·1810
16·4791	**29**	51·0342	35·7995	**63**	110·8674	54·5516	**96**	168·9408
17·0474	**30**	52·7940	36·3677	**64**	112·6272	55·1198	**97**	170·7006
17·6156	**31**	54·5538	36·9360	**65**	114·3870	55·6880	**98**	172·4604
18·1839	**32**	56·3136	37·5042	**66**	116·1468	56·2563	**99**	174·2202
18·7521	**33**	58·0734	38·0724	**67**	117·9066	56·8245	**100**	175·9800
19·3203	**34**	59·8332						

The numbers in heavy type can be either litres or imperial gallons. For example, if the heavy type number is 5, 5 litres are equivalent to 1·0999 imperial gallons and 5 imperial gallons are equivalent to 22·730 litres.

Litres-Imperial Gallons

litres		imp. gall.	litres		imp. gall.	litres		imp. gall.
4·5459631	**1**	0·21997	159·109	**35**	7·6991	309·125	**68**	14·9583
9·092	**2**	0·4400	163·655	**36**	7·9191	313·671	**69**	15·1783
13·638	**3**	0·6599	168·201	**37**	8·1391	318·217	**70**	15·3983
18·184	**4**	0·8799	172·747	**38**	8·3591	322·763	**71**	15·6182
22·730	**5**	1·0999	177·293	**39**	8·5790	327·309	**72**	15·8382
27·276	**6**	1·3199	181·839	**40**	8·7990	331·855	**73**	16·0582
31·822	**7**	1·5398	186·384	**41**	9·0190	336·401	**74**	16·2782
36·368	**8**	1·7598	190·930	**42**	9·2390	340·947	**75**	16·4981
40·914	**9**	1·9798	195·476	**43**	9·4589	345·493	**76**	16·7181
45·460	**10**	2·1998	200·022	**44**	9·6789	350·039	**77**	16·9381
50·006	**11**	2·4197	204·568	**45**	9.8989	354·585	**78**	17·1581
54·552	**12**	2·6397	208·114	**46**	10·1189	359·131	**79**	17·3780
59·098	**13**	2·8597	213·660	**47**	10·3388	363·677	**80**	17·5980
63·643	**14**	3·0797	218·206	**48**	10·5588	368·223	**81**	17·8180
68·189	**15**	3·2996	222·752	**49**	10·7788	372·769	**82**	18·0380
72·735	**16**	3·5196	227·298	**50**	10·9988	377·315	**83**	18·2579
77·281	**17**	3·7396	231·844	**51**	11·2187	381·861	**84**	18·4779
81·827	**18**	3·9596	236·390	**52**	11·4387	386·407	**85**	18·6979
86·373	**19**	4·1795	240·936	**53**	11·6587	390·953	**86**	18·9179
90·919	**20**	4·3995	245·482	**54**	11·8787	395·499	**87**	19·1379
95·465	**21**	4·6195	250·028	**55**	12·0986	400·045	**88**	19·3578
100·011	**22**	4·8395	254·574	**56**	12·3186	404·591	**89**	19·5778
104·557	**23**	5·0594	259·120	**57**	12·5386	409·137	**90**	19·7978
109·103	**24**	5·2794	263·666	**58**	12·7586	413·683	**91**	20·0178
113·649	**25**	5·4994	268·212	**59**	12·9785	418·229	**92**	20·2377
118·195	**26**	5·7194	272·758	**60**	13·1985	422·775	**93**	20·4577
122·741	**27**	5·9393	277·304	**61**	13·4185	427·321	**94**	20·6777
127·287	**28**	6·1593	281·850	**62**	13·6385	431·866	**95**	20·8977
131·833	**29**	6·3793	286·396	**63**	13·8584	436·412	**96**	21·1176
136·379	**30**	6·5993	290·942	**64**	14·0784	440·958	**97**	21·3376
140·925	**31**	6·8192	295·488	**65**	14·2984	445·504	**98**	21·5576
145·471	**32**	7·0392	300·034	**66**	14·5184	450·050	**99**	21·7776
150·017	**33**	7·2592	304·580	**67**	14·7383	454·596	**100**	21·9975
154·563	**34**	7·4792						

The numbers in heavy type can be either grams or ounces. For example, if the heavy type number is 5, 5 grams are equivalent to 0·17635 ounces and 5 ounces are equivalent to 141·7475 grams.

Grams – Ounces
(16 ounces = 1 lb (Avoirdupois))

gram		oz.	gram		oz.	gram		oz.
28·3495	**1**	0·03527	992·2325	**35**	1·23445	1927·7660	**68**	2·39836
56·6990	**2**	0·07054	1020·5820	**36**	1·26972	1956·1155	**69**	2·43363
85·0485	**3**	0·10581	1048·9315	**37**	1·30499	1984·4650	**70**	2·46890
113·3980	**4**	0·14108	1077·2810	**38**	1·34026	2012·8145	**71**	2·50417
141·7475	**5**	0·17635	1105·6305	**39**	1·37553	2041·1640	**72**	2·53944
170·0970	**6**	0·21162	1133·9800	**40**	1·41080	2069·5135	**73**	2·57471
198·4465	**7**	0·24689	1162·3295	**41**	1·44607	2097·8630	**74**	2·60998
226·7960	**8**	0·28216	1190·6790	**42**	1·48134	2126·2125	**75**	2·64525
255·1455	**9**	0·31743	1219·0285	**43**	1·51661	2154·5620	**76**	2·68052
283·4950	**10**	0·35270	1247·3780	**44**	1·55188	2182·9115	**77**	2·71579
311·8445	**11**	0·38797	1275·7275	**45**	1·58715	2211·2610	**78**	2·75106
340·1940	**12**	0·42324	1304·0770	**46**	1·62242	2239·6105	**79**	2·78633
368·5435	**13**	0·45851	1332·4265	**47**	1·65769	2267·9600	**80**	2·82160
396·8930	**14**	0·49378	1360·7760	**48**	1·69296	2296·3095	**81**	2·85687
425·2425	**15**	0·52905	1389·1255	**49**	1·72823	2324·6590	**82**	2·89214
453·5920	**16**	0·56432	1417·4750	**50**	1·76350	2353·0085	**83**	2·92741
481·9415	**17**	0·59959	1445·8245	**51**	1·79877	2381·3580	**84**	2·96268
510·2910	**18**	0·63486	1474·1740	**52**	1·83404	2409·7075	**85**	2·99795
538·6405	**19**	0·67013	1502·5235	**53**	1·86931	2438·0570	**86**	3·03322
566·9900	**20**	0·70540	1530·8730	**54**	1·90458	2466·4065	**87**	3·06849
595·3395	**21**	0·74067	1559·2225	**55**	1·93985	2494·7560	**88**	3·10376
623·6890	**22**	0·77594	1587·5720	**56**	1·97512	2523·1055	**89**	3·13903
652·0385	**23**	0·81121	1615·9215	**57**	2·01039	2551·4550	**90**	3·17430
680·3880	**24**	0·84648	1644·2710	**58**	2·04566	2579·8045	**91**	3·20957
708·7375	**25**	0·88175	1672·6205	**59**	2·08093	2608·1540	**92**	3·24484
737·0870	**26**	0·91702	1700·9700	**60**	2·11620	2636·5035	**93**	3·28011
765·4365	**27**	0·95229	1729·3195	**61**	2·15147	2664·8530	**94**	3·31538
793·7860	**28**	0·98756	1757·6690	**62**	2·18674	2693·2025	**95**	3·35065
822·1355	**29**	1·02283	1786·0185	**63**	2·22201	2721·5520	**96**	3·38592
850·4850	**30**	1·05810	1814·3680	**64**	2·25728	2749·9015	**97**	3·42119
878·8345	**31**	1·09337	1842·7175	**65**	2·29255	2778·2510	**98**	3·45646
907·1840	**32**	1·12864	1871·0670	**66**	2·32782	2806·6005	**99**	3·49173
935·5335	**33**	1·16391	1899·4165	**67**	2·36309	2834·9500	**100**	3·52700
963·8830	**34**	1·19918						

The numbers in heavy type can be either kilograms or pounds. For example, if the heavy type number is 5, 5 kilograms is equivalent to 11·023 pounds and 5 pounds are equivalent to 2·268 kilograms.

Kilograms – Pounds

kg.		lb.	kg.		lb.	kg.		lb.
0·453592	**1**	2·20462	15·876	**35**	77·162	30·844	**68**	149·914
0·907	**2**	4·409	16·329	**36**	79·366	31·298	**69**	152·119
1·361	**3**	6·614	16·783	**37**	81·571	31·751	**70**	154·324
1·814	**4**	8·818	17·237	**38**	83·776	32·205	**71**	156·528
2·268	**5**	11·023	17·690	**39**	85·980	32·659	**72**	158·733
2·722	**6**	13·228	18·144	**40**	88·185	33·112	**73**	160·937
3·175	**7**	15·432	18·597	**41**	90·390	33·566	**74**	163·142
3·629	**8**	17·637	19·051	**42**	92·594	34·019	**75**	165·347
4·082	**9**	19·842	19·504	**43**	94·799	34·473	**76**	167·551
4·536	**10**	22·046	19·958	**44**	97·003	34·927	**77**	169·756
4·990	**11**	24·251	20·412	**45**	99·208	35·380	**78**	171·961
5·443	**12**	26·455	20·865	**46**	101·413	35·834	**79**	174·165
5·897	**13**	28·660	21·139	**47**	103·617	36·287	**80**	176·370
6·350	**14**	30·865	21·772	**48**	105·822	36·741	**81**	178·574
6·804	**15**	33·069	22·226	**49**	108·026	37·195	**82**	180·779
7·257	**16**	35·274	22·680	**50**	110·231	37·648	**83**	182·984
7·711	**17**	37·479	23·133	**51**	112·436	38·102	**84**	185·188
8·165	**18**	39·683	23·587	**52**	114·640	38·555	**85**	187·393
8·618	**19**	41·888	24·040	**53**	116·845	39·009	**86**	189·598
9·072	**20**	44·092	24·494	**54**	119·050	39·463	**87**	191·802
9·525	**21**	46·297	24·948	**55**	121·254	39·916	**88**	194·007
9·979	**22**	48·502	25·401	**56**	123·459	40·370	**89**	196·211
10·433	**23**	50·706	25·855	**57**	125·663	40·823	**90**	198·416
10·886	**24**	52·911	26·308	**58**	127·868	41·277	**91**	200·621
11·340	**25**	55·116	26·762	**59**	130·073	41·731	**92**	202·825
11·793	**26**	57·320	27·216	**60**	132·277	42·184	**93**	205·030
12·247	**27**	59·525	27·669	**61**	134·482	42·638	**94**	207·235
12·701	**28**	61·729	28·123	**62**	136·687	43·091	**95**	209·439
13·154	**29**	63·934	28·576	**63**	138·891	43·545	**96**	211·644
13·608	**30**	66·139	29·030	**64**	141·096	43·999	**97**	213·848
14·061	**31**	68·343	29·484	**65**	143·300	44·452	**98**	216·053
14·515	**32**	70·548	29·937	**66**	145·505	44·906	**99**	218·258
14·969	**33**	72·753	30·391	**67**	147·710	45·359	**100**	220·462
15·422	**34**	74·957						

Appendix 2

pH – Hydrogen Ion Concentration

This is the measure of the number of grams of hydrogen ions per litre of solution, i.e. the actual number per unit volume and is the way acidity or alkalinity is measured.

When the pH is 7 we have a neutral solution, i.e. water. Below this from 1–7 we have acid solutions, whilst above we have alkaline. These relationships are best illustrated by reference to the pH scale shown below:

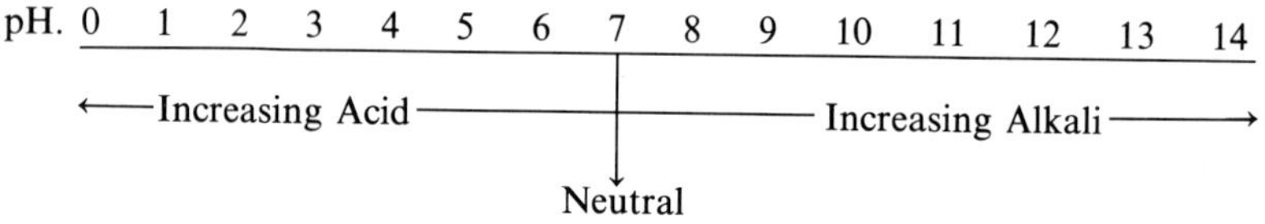

Following this diagram it can be seen that as we move down the scale to 0 we get increasing acid concentrations, whilst when we move towards 14 increasing alkaline solutions are obtained.

Thus an acid of pH6 is weaker than one of pH5 and conversely an alkali of pH8 is weaker than one of pH9.

Index

ANTHONY SMITH